I0794302

This book graciously and simply explains why and how evolution is a critically important factor for understanding the development of Earth's life, but why it's not the only factor. Sy Garte explains why, in many instances of the origin and history of life, the divine hand of God creates, designs, and guides. Sy shows how biology will most productively advance with a both-and, not an either-or, approach. I highly endorse this book and believe it should be required reading for every biology student and church leader.

**HUGH ROSS, PhD,** pastor, astrophysicist, and founder and senior scholar of Reasons to Believe

The Berlin Wall of evolution fell, but there were no reports in the media. The iron curtain between faith and science has lifted. Sy's book heralds the new era.

**PERRY MARSHALL,** author of *Evolution 2.0*

In this thoughtful and well-written book the author describes how concepts such as "purpose," a word long banned in evolutionary discourse, have now come back into the biological mainstream. Delving into the biggest questions that biology yet has to resolve, Sy Garte sees God's creative work in the whole of life, from beginning to end, not somehow located in the current gaps in our scientific understanding. Those of any faith or none will find much here to stimulate thought and to inspire.

**DENIS ALEXANDER, PhD,** emeritus fellow of St Edmund's College, Cambridge University, and emeritus director of The Faraday Institute for Science and Religion

A devout Christian and a highly accomplished biochemist, Sy Garte has become one of the most important voices in the science-faith conversation today. And one of the more interesting. And he doesn't disappoint in his latest work. In my view, it is imperative for anyone interested in the science-faith conversation, particularly for those of us who barter in the life sciences, to carefully consider Garte's powerful points in *Beyond Evolution*. Garte challenges both Christian and atheistic biologists to reconsider biological evolution in the face of emerging insights about the nature of the evolutionary process. Doing so just may turn your thinking about the living world upside down. Garte points out that a revolution is underway in biology with long-forbidden concepts—such as teleology, agency, and cognition—pushing their way to the forefront. According to Garte, these insights mean that biological evolution should no longer be considered a threat to the faith of Christians and should no longer serve as the basis for atheists to dismiss a Creator's role in life's origin and history.

**FAZALE "FUZ" RANA, PhD,** biochemist and president and CEO of Reasons to Believe

For too long, biologists have tried to reduce life to chemistry and physics. Garte exposes the futility of that project and makes their dead biology come to life. He shows how purpose, goals, agency, choice—and even love—are characteristics of living things that cannot be reduced away. If created reality at its most basic is just atoms and the void, it's not difficult to see why atheism is attractive to some people. But in Garte's inventory of the fundamental constituents of the universe, life and its capacities are counted among them. That doesn't prove their origin is an Author and Giver of Life, but it certainly points us in that direction.

**JIM STUMP, PhD,** author of *The Sacred Chain*, vice president of BioLogos, and host of *The Language of God* podcast

In *Beyond Evolution*, Garte presents a compelling case for a fundamental teleology in the biological realm. Regardless of one's position on evolutionary theory or the biblical creation narrative, there is much of value to be gleaned from this concise volume. It is a fascinating and insightful contribution to the science and faith conversation.

**MELISSA CAIN TRAVIS, PhD,** fellow at Discovery Institute's Center for Science and Culture

*Beyond Evolution* is a watershed book that widens and deepens the conversation on the science of biology and its relationship to belief in God. Many see these two as having a more checkered relationship than, say, for the physical sciences. But Garte encourages us that it need not be so, and that new developments in the field of biology point ever more strongly to a "divine creator." *Beyond Evolution* is expertly written and beautifully crafted. I highly recommend that you read it.

**SHARON DIRCKX, PhD,** speaker, broadcaster, former neuroscientist, and author of *Am I Just My Brain?*

# BEYOND EVOLUTION

# BEYOND EVOLUTION

*How New Discoveries in the Science of Life Point to God*

SY GARTE, PHD

TYNDALE
REFRESH®
*Think Well. Live Well. Be Well.*

Visit Tyndale online at tyndale.com.

*Tyndale*, Tyndale's quill logo, *Tyndale Refresh*, and the Tyndale Refresh logo are registered trademarks of Tyndale House Ministries. Tyndale Refresh is a nonfiction imprint of Tyndale House Publishers, Carol Stream, Illinois.

*Beyond Evolution: How New Discoveries in the Science of Life Point to God*

Copyright © 2025 by Seymour Garte. All rights reserved.

Cover and interior images are the property of their respective copyright holders from Adobe Stock, and all rights are reserved. Spiral leaves © 418studio; fractal © Digital-Art; nautilus shell © barneyboogles; golden ratio © O_a.

Author photo by Cameron Bertuzzi, copyright © 2021. All rights reserved.

Designed by Eva M. Winters

Edited by Jonathan Schindler

Published in association with the literary agency of The Steve Laube Agency.

All Scripture quotations, unless otherwise indicated, are taken from the Holy Bible, *New International Version*,® *NIV*.® Copyright © 1973, 1978, 1984, 2011 by Biblica, Inc.® Used by permission. All rights reserved worldwide.

Scripture quotations marked ESV are from The ESV® Bible (The Holy Bible, English Standard Version®), copyright © 2001 by Crossway, a publishing ministry of Good News Publishers. Used by permission. All rights reserved.

The URLs in this book were verified prior to publication. The publisher is not responsible for content in the links, links that have expired, or websites that have changed ownership after that time.

For information about special discounts for bulk purchases, please contact Tyndale House Publishers at csresponse@tyndale.com, or call 1-855-277-9400.

**Library of Congress Cataloging-in-Publication Data**

A catalog record for this book is available from the Library of Congress.

ISBN 979-8-4005-0136-4

Printed in the United States of America

31 30 29 28 27 26 25
7 6 5 4 3 2

*This book is dedicated to my loving wife, Anikó Albert.*

*For his invisible attributes, namely, his eternal power and divine nature, have been clearly perceived, ever since the creation of the world, in the things that have been made. So they are without excuse.*

ROMANS 1:20, ESV

# Contents

# Foreword

IN HIS INTRODUCTION TO THIS BOOK, Sy Garte says that doing good science requires courage. This strikes me as an apt description of the scientific enterprise.

After all, science is about asking hard questions and being willing to revise our understanding of the world based on new evidence revealed by our experiments. That requires courage, especially when new findings challenge a much-loved theory or cherished consensus. And especially when the evidence points in the direction of purpose, design, or even God.

Until recently, God has been banished from view in most scientific circles as a materialist orthodoxy has taken over the academy. When biologist Richard Dawkins proclaimed in his book *The Blind Watchmaker* that "Darwin made it possible to be an intellectually fulfilled atheist," it reflected the fact that he and many other scientists regard a naturalistic theory of evolution as a complete explanation for life in all its complexity and variety. God, they believed, was no longer necessary in a world explained by unguided physical processes acting over long time periods. Indeed, as a completely materialist view of the world grew in popularity, "scientism"—the view that everything can be explained in purely natural, scientific terms—became a substitute for God in the hearts and minds of many.

As the host of many faith debates over the past twenty years, I have been witness to the dogmatic insistence of those in the scientific community who refuse to countenance any explanations that will admit a divine "foot in the door." This has led to conversations with some notable scientific figures who, when asked if there is *any* evidence that might point in the direction of God, have insisted there is not.

"But what if the stars lined up in the sky to spell 'It's God . . . believe in me!'?"

"It could be advanced alien technology."

"What if Jesus appeared to you in this room right now?"

"I would assume I was having a brain seizure."

And so on.

It turns out that, rather than being paragons of objectivity and reason, scientists are as human as the rest of us. Indeed, there are some scientists who have become so wedded to their preexisting worldview that it appears no amount of evidence would change their mind.

Happily, that is not the case with Sy Garte. His personal journey from skeptical atheism to Christian faith in late middle age is a testimony to the fact that he is willing to follow where the evidence leads. The same applies to Sy's intellectual life as a scientist.

Written with the layperson in mind, *Beyond Evolution* is an indispensable guide to how fast the landscape of the natural sciences is changing, as a paradigm shift in physics, chemistry, and biology takes place. The development of life, it turns out, is not the result of a mechanistic, unguided process but rather exhibits an extraordinary precision and purpose that points to something deeper and more mysterious than mere physics and chemistry.

Alongside Sy, many significant scientists have broken loose from the shackles of materialist orthodoxy and are courageously offering a revolutionary account of cosmology and biology that overturns the materialist assumptions of many of their predecessors. As the notable physicist and origins researcher Professor Paul Davies, an agnostic, put it to me, "The directionality of the Universe going from matter, to life, to consciousness . . . to comprehension. That comes very close to something like a meaning or purpose in nature. I think that's a sort of religious feeling. What Einstein called 'a cosmic religious feeling.'"

Whether you call them dissidents, pioneers, or something else, Paul Davies, Sy Garte, and many others are in the scientific vanguard of a movement that I have dubbed "the surprising rebirth of belief in God."

The expulsion of God from the realm of science was premature. Indeed, it may be that not only does science point to God, but that God has been responsible for our ability to do science all along.

I am glad for the courageous voice of Sy Garte. In reading this book, I am confident that you will be grateful for it too.

*Justin Brierley*

Host of *The Surprising Rebirth of Belief in God* podcast
and author of *Why I'm Still a Christian*

INTRODUCTION

# How Did We Get Here?

FINDING THE RIGHT QUESTIONS to ask is the key to progress in all kinds of endeavors[1]—from science to plumbing to theology. Probably one of the oldest questions people have asked themselves is "How did we get here?" The word *we* in this question can mean us humans, all of life, or even all of existence. The question has clear scientific as well as theological aspects, and in some cases (especially when things go wrong) can even apply to plumbing.

This book is about how the science of biology and our understanding of the majesty of God can answer that question applied to life on earth. You will read about how evolutionary theory is being transformed and how single cells do things like make decisions and communicate with each other. I discuss the origin of life and of human consciousness, the reality of purpose and agency in the science of life, and why we need new laws in biology to move the science forward. Above all, this book is about the strong connection between life and divine creation, culminating in the existence of you and me, a human reader and a human writer of these words, creatures made in the image of God.

The overarching message I hope to convey is that modern mainstream biological research in several fields is now providing evidence for a divine creator. As Ruth Bancewicz of the Faraday Institute puts it, "The question 'Can Science Prove God Exists?' turns out to be a category error—trying to use science to answer a non-scientific question. On the other hand, many scientists think there is enough evidence from a wide range of sources to warrant belief in God."[2]

I am among those many scientists. And in this book, I will try to convince you to join me. If God exists and is the creator of the world we know—meaning all of nature—it is logical to assume that the natural world will contain signs of that divine creation. If there were no God, then the entire universe would conform to Richard Dawkins's description:

> The universe we observe has precisely the properties we should expect if there is, at bottom, no design, no purpose, no evil and no good, nothing but blind, pitiless indifference.[3]

However, the fact that there is love, beauty, purpose, and meaning in at least one part of the universe—here on earth, inhabited by loving, purposeful people who create beauty—counts as evidence for the existence and majesty of a divine presence. This is what the early empirical scientists (most of whom were devout Christians) meant when they based their work on finding the God-given laws that govern the natural world. Creation declares the glory of God, the Good Book tells us. As a biochemist, I see God's glory most clearly in biology.

To correct Dawkins, we can say,

> The universe we observe has precisely the properties we should expect if there is, at bottom, a grand design, a purpose (such as the existence of life that can worship its

Creator), good and evil, and all that is needed to justify belief in the presence of God.

In the next two chapters, we'll apply the central question "How did we get here?" to the current rift between differing groups of Christians and biologists. How did the science of life become a battleground for so many Christians? How has evolution become the "theory of everything" that atheistic scientists use to explain a universe without God? I will emphasize that we need to move beyond the "evolution wars" that divide Christians and beyond a purely reductionist, materialist view of life that ignores so many new advances in science, including a transformation of evolutionary theory itself.

The reality is that the theory of evolution is neither atheistic nor antibiblical—it is fully consistent with the existence and omnipotence of a divine creator. Furthermore, contrary to common belief, evolution by natural selection is not the most important principle of the life sciences but the result of more fundamental, uniquely biological processes. I will show you why the time has come to stop fighting over the reality of evolution among Christians.

We are at the beginning of a scientific revolution—perhaps a paradigm shift—in biology that will replace concepts such as random chance and blind selection as the major drivers of biological change with new approaches encompassing purpose, agency, and cognition in cell biology.

The origin of life is the greatest mystery of biological science and will not be understood using only the known laws of chemical or physical science. This mystery can only be approached by brand-new, biology-specific theories.

And finally, we must ask: "How did we human beings get here?" The enormous complexities of human behavior and consciousness serve as pointers to the concept of being made in the image of God.

## The Paradigm Shift

Central to my message is the need for new scientific laws specific to biology. These include evolution, which at present has no mathematical law to bolster the theory (chapter 2); the origin of life, which doesn't even have a theory (chapter 5); and the nature and theory of consciousness, which has neither law nor theory—or even a good definition (chapter 6). Potential sources for some of these laws are given in chapters 3–5. I have no doubt that once we have discovered these new laws, specific to the science of biology, they will point to the designer God as their source, as do all the laws of nature.

In order to accomplish this, biology is in dire need of accepting some concepts that will strike many biologists as scientific heresy. We need to revise a great deal of our current theories to include things like teleology, agency, cognition, and many non-gene-centric ideas of how living creatures operate and survive against all odds. I will reveal some exciting recent research that casts serious doubt on the widely accepted premise that biological design is completely devoid of guidance by willful agency.

There was a time when the clockwork mechanisms of classical physics seemed entirely deterministic and materialist, with no hint of mystery and, at least the way some people saw it, no need for God. That time has long passed as the introduction of quantum theory, relativity, the fine-tuning of the cosmological constants, and the evidence that the universe had a beginning have reshaped so much of modern physics.[4]

Signs of this kind of shift are starting to appear in biology with data and experimental evidence pointing to something beyond our current understanding. The theories that will become established to explain these new findings are yet to be formulated, but their time is coming. As a matter of faith in God and in the scientific process, I believe that, as always, the discovery of these putative new theories and laws governing life will point to the truth of a divine creator.

## Biology Is Different

The study of life is a science, of course—the science of biology. But life is different from planets, rocks, and molecules, and thus the science of biology is also different from the other sciences. This is because many of the things that life can do—like reproduce itself—are not found in any other kind of matter. Exploring those differences will be a central part of this book.

I remember sitting as a young man in a graduate class in biochemistry and being utterly astonished by what I was learning about the details of how proteins are made in cells. And when I taught it as a professor, I remember seeing the same look of wonder on my students' faces and knew that I had succeeded as a teacher.

Hoping to reproduce that effect, I will try my best here to describe what I often call the incredible complexity of biochemical processes, including how cells reproduce themselves. I've tried to keep most of this explanation in plain language, but for those who would like to know more, I include a technical appendix that goes into greater detail regarding the mechanisms of protein synthesis, the genetic code, and other biochemical machinery in cells.

Biological research is opening the door to a broader and deeper understanding of the foundational principles underlying the science of life. These principles also have theological ramifications that go in the opposite direction from those of reductionist ideology.

Most professional biologists (including myself in the past) see Darwin's theory of evolution by natural selection as the biological "theory of everything." While not denying the important role evolution plays in life, I intend to convince you in chapter 3 that the most fundamental principle of biology is not evolution but accurate self-replication leading to inheritance. I will show you how the process of evolution is simply an automatic outcome of inheritance by self-replication and that replication is both necessary and sufficient for evolution by natural selection to happen.

*Biology is different* is a theme I elaborate on throughout the book as I discuss the role of natural laws in the other sciences and the lack of such laws in biology. I argue that in order to go further in our understanding of life at its deepest levels, we will need to use ingenuity and courage to incorporate previously forbidden concepts like purpose, agency, and primitive cognition into our explanatory paradigms and theories. As examples, I bring in many newer scientific ideas published in peer-reviewed scientific literature that have not yet reached popular culture.

You might think "courage" is a strange word to use in a scientific context, but there is precedent for its application in advancing a field of science to a new level of understanding. The courage of Albert Einstein to propose a radical new theory of the nature of light, involving such seemingly bizarre concepts as time slowing down as a function of speed, brought physics into the modern era and led to a better comprehension of the mysteries of time, space, matter, and energy.[5]

But even Einstein, the formerly bold pioneer, would later balk at the notions of probabilistic reality revealed by quantum theory. In order to make the kind of progress that the new physics led to (giving us smartphones and GPS, MRI machines, and nuclear energy), the older philosophical doctrines of reductionism and strict determinism had to be laid aside. In this book, I discuss these same doctrines and their stifling effect on biological progress and highlight the equivalent work of courageous modern-day biological scientists who are emulating the bold spirit of Einstein, Planck, Bohr, and Heisenberg.

## Biology and Theology

The questions "How did we get here?" and "Where are we going?" are connected, and that connection is, I believe, being brought into ever sharper focus by new discoveries in molecular biology, evolution, and other fields of biological science. Ultimately, I believe the complete truth about our world and ourselves requires input from theology as well as science. There is only one universal truth, and in order to discover it, we need both laboratories and churches, Scripture and equations, worship and peer review.

If you have read my book *The Works of His Hands: A Scientist's Journey from Atheism to Faith*, you know that I came to faith in late middle age, and I have little experience or knowledge of apologetics. I'm still learning some details of Christian life that are second nature to cradle Christians. Yet I know that science, especially the science of biology, is strongly harmonious with Christianity (see my book *Science and Faith in Harmony: Contemplations on a Distilled Doxology*[6]).

There really is something about life beyond the laws of thermodynamics and chemical equilibrium. There is, not just in us but in all of biology, a divine spark. I will show you the evidence that what I just said is true and how that truth fits in with the holy truth of Christ and the Bible. We humans are indeed made in the image of God, and all of life is imbued with the spirit of its creator.

Not being a theologian, I rely on the work of others when it comes to questions about how scientific knowledge relates to biblical truth. As scholars in the science-and-faith community often emphasize, it's a mistake to directly compare our scientific knowledge with the writings of Genesis and other books of the Bible. It makes no sense to read the Bible as a scientific textbook. It would in fact be foolish to read even an eighty-year-old science textbook as a source of accurate modern scientific information, let alone trying to fit a series of books written thousands of years ago about the nature of God and humanity into any kind of modern scientific narrative. The Bible's purpose is to teach theological truth—to tell us about the Creator that loves his creatures and wants a relationship with them.

Justin Brierley writes in his book *The Surprising Rebirth of Belief in God*,

> Evidence from history, philosophy, theology, personal experience, and the testimony of others will all play their part in drawing thinking men and women towards Christian faith. But for many, an awed wonder at the complexity of the very smallest building blocks of life—and at the very largest aspects of our cosmos—may be the place where such journeys begin.[7]

I do believe that once properly understood, biology—including evolution in its modern and ever-improving versions, as well as molecular biology, ecology, human consciousness, and the origin of life itself—is a strong scientific pointer to the majesty of the divine creator. This book is all about demonstrating the evidence for that belief to believers and skeptics. I hope reading it will help biologists, students of biology, and the general public see the world of life in a new light, even in the light of Christ. We will see the view of nature as pitiless and indifferent transformed into a brilliant light: one of conscious design, mighty purpose, and overwhelming love. And we will (as I did as a mature adult) come to know that the study of life is part of the universal praise of God, pure and refined. For biology, like all the natural sciences, is distilled doxology.

We will start with a discussion of the historical context in which biology originated and developed into a real science, as well as how we got to the pointless and destructive bitterness related to the theory of evolution within the church. We will then tackle the subject of why evolution is not worth fighting about. Let us now embark on a wondrous journey into a new world of biological reality that does nothing less than declare the glory of God.

CHAPTER 1

# Biological Ideology and a Call for Unity

We often hear the phrase "we live in troubled times." It is not a very profound statement—it has always been true, although the kind of trouble has varied. For Christians, part of the current malaise is a bitter antagonism within the church over biblical interpretation and the role of science. While there have always been divisions among followers of the Way, in the present case, I believe at least some of the disputes related to evolution are resolvable, and I will discuss this in the latter part of this chapter. But first let's take a close look at how biology got where it is today.

I believe that nothing is ever thoroughly comprehensible without a knowledge of its history. A good deal of what I want to say in this book about biology might not make much sense without an understanding of how the science of life developed, and how this history has influenced modern ideas about how to study life scientifically.

Biology is the youngest of the traditional natural sciences, and some branches (population genetics, systems biology, molecular biology) have only been around for a century or less. It is an exciting science—not only because it is the science of the most beautiful, God-given phenomena of life, which includes us humans, but because of its freshness, its plethora of new frontiers, its host of unanswered questions, and its potential for illumination of the magnificence of divine action in the natural world.

It is my firm contention that progress in biology, while undeniably swift and exponentially successful, is also hampered by some ideological attitudes and outdated dogmas. Some of these are currently in the process of being reexamined and rejected, which is very good news. To provide a foundation for the discussion of that good news in the rest of this book, I will highlight here the history of how biology came to be what it is today.

## Origins of Modern Science

Modern empirical science arose in medieval Europe when several pathways of thought coalesced into a brand-new worldview. The "natural philosophers," inspired by their Christian faith, understood nature as God's creation and thus orderly and lawful and able to be studied and understood. In general, the church supported this fundamentally monotheistic idea.

Another critical thread in the development of science as we know it was the application of mathematics to the discovery of new natural laws. Thinkers marveled at the fact that the basic workings of the natural world could be understood and written down as simple equations. Isaac Newton revolutionized the worlds of science and philosophy with his demonstration that the same force that makes an apple fall to the ground also governs the motions of the planets.[1] The unifying principle of gravitation showed that all of reality is connected both in time and space—the universe is truly governed by laws. Newton saw this as a demonstration of the work of the divine lawgiver, as did his colleagues for many centuries.

As natural philosophy embraced empiricism, inductive reasoning, and the power of experimental methods, what we now call the scientific revolution unfolded. It wasn't long before the world began to see the fruits of these concepts in many technological breakthroughs.

## Physics and Metaphysics

There also grew a tacit understanding that the realm of science was physics and the realm of theology was metaphysics. The two agreed not to interfere in each other's realms, though the definition of what was physics and what was metaphysics differed from how we understand those terms today. The nature of consciousness, for example, was not a subject for science, nor was any deep investigation into the nature and origin of life. Life, and human life in particular, continued to fall under the biblical and theological purview.

On the other hand, discoveries related to chemistry, research on thermodynamics, and the intricate mechanical inventions of the time provoked nothing but admiration and respect from everyone, clerics included. By the end of the eighteenth century, science was increasingly hailed as the new wisdom and a source of unlimited potential.

As the new discoveries continued, the idea that all of nature was within the intellectual grasp of science started to grow and gain consensus in the second half of the nineteenth century. Maxwell's laws showing the close relationship of electricity and magnetism were the crowning touch,[2] and many began to believe that the dream of a full human understanding of nature was within reach.

Naturalism was triumphant, and there were models to explain everything. Light, for example, was understood to be a wave that travels in the ether, atoms were particles that obey the laws of classical physics, living cells were thought to be composed of a homogeneous jellylike substance called protoplasm, and the universe was held to be past-eternal (meaning it has always existed) and in a steady state.

## The End of Certainty

The dawn of the twentieth century, however, burst the bubble of scientific rationalism championed by so many enthusiastic thinkers. Einstein proposed theories, confirmed by experiments, that fundamentally changed the way we think about the nature of time, space, and gravity. The subsequent revolution in quantum physics destroyed the notion of scale independence of the physical laws: what was true for planets and apples was not at all true when extrapolated down to the subatomic world. It turned out that the unity of physics in controlling the behavior of objects did have a size limit after all.

We can thus describe the history of the physical sciences in two major phases, at least so far. The first phase, the growth of certainty, was the enlightened view that with observations and experiments, the complete truth about the world could be obtained. This was a period of solving mysteries, closing knowledge gaps, and forging ahead with confidence that science would leave no question unanswered.

The second phase, the rise of uncertainty, began in 1905 with Einstein's publication of several papers, including one on the theory of special relativity.[3] Throughout the twentieth century, spearheaded by developments in quantum mechanics, fractals, nonlinear dynamics, and chaos and complexity theories, the second phase was marked by a questioning of the reductionist paradigm that was the basis of the first phase.

While the first phase was reductionist and optimistic, the second phase was neither. Heisenberg's uncertainty principle (1927), which states that it is impossible to know both the momentum and position of an electron at the same time,[4] undermines the idea that science can answer all questions. Soon after, Gödel's incompleteness theorem (1931) showed that even mathematics—the pure language of science—was not immune to uncertainty.[5] In this phase (which continues today), many questions have been answered, but the answers have always led to more questions. Reductionism and pure determinism increasingly became nonviable. Mysteries seemed to proliferate, and the old confidence that science would soon solve all of them became a philosophical view held by ideologues but very few scientists.

## How Can We Trust Science?

If scientific certainty is a thing of the past, and science keeps changing, it's worth asking: How can we trust what scientists say? How do we know that they won't change their tune when new information comes along? That does, in fact, happen, and it simply means that rarely are we absolutely sure of scientific truth. But science is based on honesty, and scientists tend to admit freely when they make a mistake or when some understanding changes—and when they don't, their peers will hold them to account.

How does that process work? Scientists communicate with each other by writing scientific papers (not news articles) that are published in scientific journals (not magazines, books, or blog posts). The way scientific journals decide whether or not to publish a scientific paper is to send the manuscript to two or three other scientists in the same field and ask them to evaluate it.

This process is called "peer review." The people chosen to be peer reviewers have generally reached a certain level of recognition and experience, and they have a reputation for honesty and integrity.

Since there is no peer review to check the accuracy and quality of the information contained in blogs, social media, newspapers, magazines, short reels, YouTube videos, TV news clips, and so on, we cannot be sure that such information is either thorough or accurate.

## The Origin of the Science of Biology and the Darwinian Revolution

While the physical sciences were going through their periods of certainty and uncertainty, the science of biology was on a different course.

When Charles Darwin began his studies, biology had already been divided along the lines of orthodoxy. Some felt that all work in biology had to conform to religious scripture, since life was held to be squarely within the realm of the metaphysical. Others, such as most of Darwin's friends and teachers, held the view that biology should become a science much like chemistry and physics. This meant that the rules of science as they began to be understood for physics and chemistry would apply to scientific inquiries on the nature and detailed mechanisms of life.

The movement to make biology a science was not antireligious. It was a project by a group of thinkers who wanted to apply the methods of the successful sciences to understanding nature better. The boundaries of science were evolving, and moving life out of metaphysics and into the scientific category was a logical next step.

Carl Linnaeus, Jean-Baptiste Lamarck, and others began this trend where it needed to, with classification schemes of living beings and fundamental ideas about whether and how organisms could change and emerge.[6] Other scientists, most of them originally chemists, entered the field of biology with experimental and theoretical advances. The chemistry of living creatures became a subject for study and analysis, just as chemistry itself was making huge strides in technical and theoretical progress. The work of Louis Pasteur, Claude Bernard, and others gave hope and encouragement to those looking for a new, rigorous science of life. The term *biochemistry* entered the scientific lexicon in 1903, and discoveries followed at breathtaking speed.

But there was something important missing: an idea that could form a fundamental basis for biological science the way Newton's theory of gravity unified physics.

When Charles Darwin published *On the Origin of Species* in 1859, it shook up the way people thought about life but did not lead to an immediate revolution in biology. Darwin's theory of evolution was at first considered an interesting intellectual exercise, a useful description of how biological diversity might arise. But the fact that it had no foundation in chemistry, physics, or mathematics led to a gradual loss of interest among both scientists and the public as the nineteenth century ended. Meanwhile, biologists were moving on to laboratory studies of proteins, enzymes, membranes, and other fascinating components of living cells.

That all changed at the turn of the century with the rediscovery of Augustinian friar Gregor Mendel's work on the laws of heredity, leading to the birth of genetics. For the theory of evolution by natural selection to work, Darwin had assumed, based on common observations, that organisms reliably passed on their characteristics to their offspring, but he had no understanding of how they did so. Now the missing details were being worked out, and the scientific basis of heredity was firmly demonstrated. Darwin's theory, built on the universal techniques of plant and animal breeding and on the bedrock of Malthusian population dynamics, began to regain adherents among biologists.

When R. A. Fisher, J. B. S. Haldane, Julian Huxley, and others put the theory together with the genetic theories of Mendel, the modern synthesis of evolutionary biology soon became the central tenet of all biological science. Vast amounts of new information have been generated since then, spanning paleontology to molecular genetics, cell biology, animal behavior, and ecology, encompassing every branch of biological science, all of it in accord with the basic idea of evolution by natural selection.

## The Neo-Darwinian (Modern) Synthesis

The twentieth century saw the flowering of Mendelian inheritance into a major science. The concept of the gene—the hypothetical inheritance particle—was substantiated by the discovery of the structure and function of DNA.

The phenomenon of mutation—a change in the function of a gene—was understood even before anything was known about the chemical nature of these theoretical controllers of inheritance. Mutations could be studied by their effects on the characteristics of fruit flies or bacteria growing in dishes. One important question was the role of chance as opposed to purpose in causing mutations. Can organisms purposely act to produce mutations in specific genes that would help them survive an environmental stress, such as starvation or exposure to toxic agents? While this idea seemed to make sense, many believed that an alternative scenario, which fit Darwinian theory more closely, was correct. In this hypothesis, mutations are generated randomly, and then the lucky few that had a beneficial mutation survived due to the principle of natural selection.

In the 1940s, experiments in bacteria seemed to definitively answer this question. The results showed that mutations were produced entirely randomly, and natural selection determined which cells would survive based on what genotypes led to the phenotypes that did best in the given environment.

As we will see in later chapters, this conclusion has been shaken by recent research, but at the time, these results helped to confirm the emerging neo-Darwinian view that purpose plays no role in the mechanisms of molecular genetics. The triumph of blind chance as the key factor in the first stage of evolution was absorbed into biological dogma, and, as in the rest of natural science, teleological ideas were discarded from consideration. Evolution became a theory devoid of any hint of purpose or design.

## DNA and the Molecular Revolution

The experimental findings of Oswald Avery in the 1940s that DNA might be the chemical responsible for heredity or the "genetic material"[7] were initially met with astonishment and doubt. The expectation had been that the answer to the chemical mystery of the nature of genes would be a protein enzyme. DNA, with its limited chemical composition, didn't seem interesting enough to be a good candidate, considering the complexities of inheritance. The discovery of the double helical structure of DNA by James Watson and Francis Crick in 1953[8] changed all that, and the revolution in the new field of molecular biology was on.

In the decades that followed, an unprecedented crescendo of discovery revealed how cells work on a molecular level, including how they inherit their characteristics and how they do all that they do to remain alive.

Progress in molecular biology got another boost in 1970 with the discovery by Howard Temin and David Baltimore of reverse transcription in viruses and the technical possibility of recombinant DNA and cloning.[9] This led inevitably to one of the great projects in the history of science—the sequencing of the entire human genome by the Human Genome Project,[10] completed in 2003.

## Twenty-First-Century Biology

As we move through the first decades of the newest century, there are signs that biology might be entering the second phase of scientific discovery, similar to what physics experienced a century earlier. Increasingly, the answers to research questions are leading to many surprising answers and to new, more difficult mysteries.

One of these surprises has been the finding from the Human Genome Project that humans do not have a much larger collection of genes than most other organisms. It appears that the reason we are such complex biological beings is not because we have more genes than other creatures but because we regulate their expression in highly complex and interesting ways. This means that the control of *when* a gene does something is just as important as *what* the gene does.

Progress in biology has included new techniques like gene cloning (enabling genes to be copied and studied in detail) and gene editing (enabling changes to be made to the sequence of genes to correct mutations or alter their functionality).

Another field that has grown exponentially is systems biology, which, thanks to advances in computer modeling, is beginning to allow researchers to make sense of the millions of interactions and interwoven biochemistry of even the simplest cells.

Having briefly covered the general history of biology from its origins up to the present, let us have a closer look at parts of that history that are most important to my message in this book: the history of the relationship between science and Christian faith, and the history of biological ideology, apart from scientific progress.

## The History of the Relationship Between Science and Faith

There is no dispute that most of the major European scientists from the origin of science up to the end of the nineteenth century were Christians. But of course, they lived in a Christian milieu and might not have had much choice—at least, that is what we hear from the kind of skeptics who insist on the basic incompatibility between a scientific and a Christian worldview. Their view of the relationship between science and faith has come to be known as the "conflict thesis" and has its origins in the same late-nineteenth- to early-twentieth-century period that saw so many intellectual, technological, political, and social revolutions at once.

It must be pointed out that most of those who accused faith of being an enemy of science were not actually scientists (as is often true today). So how did the conflict thesis arise in the first place? As always, there are many reasons, covered extensively by historians, but the immediate catalyst was the publication of two books. The first was *History of the Conflict Between Religion and Science* by chemist and inventor John William Draper, followed about twenty years later by Andrew Dickson White's *A History of the Warfare of Science with Theology in Christendom*.[11] White was not a scientist or a theologian but a historian, politician, and educator. These books were reacting to attacks on Darwin's theory of evolution by some clergymen, but they generalized this conflict to all of history to paint a picture in which theologians and the church had always stood in the way of progress, oppressing science and scientists. Their false narrative had an enormous influence on educated elites in the United States and Europe, leading to a major shift in public thinking about the relationship between science and faith. But the tide might now be turning, as reported in a recent book by David Hutchings and James C. Ungureanu:

> Thanks to the dedicated and committed research of a band of specialists operating since the 1980s at least, the conflict thesis has now been thoroughly debunked. One by one, the tales spun out in *Conflict* and *Warfare* have been shown to be either entirely false, horribly misunderstood, or deliberately misrepresented.[12]

The popularity of the conflict thesis likely benefited from social changes starting in the same period the two books were written. The explosive development and increasing importance of new technology in Western countries began to sweep away traditional lifestyles and, with them, more spiritually centered worldviews for large swaths of the population. The idea of an ongoing conflict between religion and science became ingrained in the culture with many other ideas of modernity, despite its harmfulness to both. This harm has arguably been greatest in the United States, where the majority of the population remained Christian and where these misguided assumptions of a war between science and faith have been used to divide people along political and culture-war lines.

Building in part on the same narrative, the rise of New Atheism at the beginning of the twenty-first century initiated a wave of deconversion in the developed world. Younger, educated people bought into the rehashing of the conflict thesis by the likes of Richard Dawkins and Sam Harris. Many scientists as well as philosophers, entertainers, and public figures joined the movement toward atheism.

## The History of Ideology in Biology

There is no question that the rise of biology as a true science was difficult. The phrase "physics envy" has been used to highlight the somewhat defensive posture of biologists who are said to suffer from an inferiority complex due to the lack of mathematical rigor and clear structure of biological information and theory. In response, biologists went through a period of purging from their field any concepts that could possibly be construed as "nonscientific," meaning not fitting well with the reductionist nature of pre-twentieth-century physics and chemistry. This defensive reaction initially served biology well but with time has prevented progress by limiting theoretical explorations to only those that pass the test of strict methodological naturalism. Concepts like purpose or agency were seen as the proverbial camel's nose, which, if allowed into the tent of pure science, would soon be followed by the rest of the camel—all kinds of nonscientific metaphysical ideas.

The first such concept to be banished was *vitalism*, the belief that living things possess something more than inanimate matter. In the eighteenth and early nineteenth centuries, as physics and chemistry grew to explain more and more of the world of matter, many thinkers and scientists felt that life was animated by a "life force" or "vital principle" that could not be understood or probed with microscopes, dissections, and experiments. (The *mechanists*, those who thought living things were just complex machines, disagreed.) Some vitalists proposed that even the chemistry of living beings was fundamentally different from abiotic chemistry and that the "organic" chemicals of life could not be synthesized in a laboratory. In their most extreme version, such beliefs made the idea of biology as a natural science impossible to consider. As a result, biologists increasingly considered vitalism to be a toxic concept.

When urea, a simple organic chemical compound found in all mammals, was synthesized in a lab in 1828, one of the assumptions of strong vitalism was clearly shown to be false: components of life *could* be duplicated outside a living organism, without the presence of a special vital force. Vitalism gradually faded both from philosophy and science, allowing for an unhindered beginning of a new science of life that used all the tools that chemical and physical sciences had to offer.

The semimystical concept of vitalism was not the only victim of the desire to purge what were considered nonscientific concepts from biology. One of the hallmarks of modern scientific methodology in biology is the removal of teleology, or purpose, from scientific consideration. The idea that there is no teleology in science has worked well for physics, chemistry, geology, and a great deal of biology as well. Nothing in the world of physics or chemistry acts with purpose—all that happens is strictly due to the laws of nature, which also include random chance. No volcano decides to erupt and then gathers the energy and materials to do so. Oceans don't consider the idea of being tidal—it's not up to them but to the gravitational forces from the moon.

With early advances in molecular biology, including the apparent random nature of mutations and the chemically driven mechanisms of physiology, it certainly seemed that biology had no need or use for teleological thinking either. However, it is hard to deny that most animals, especially we humans, do in fact act with purpose—or at least the appearance of purpose and the appearance of free will, as the strict reductionists would say. But recent scientific discoveries, discussed in detail in chapter 4, have begun to paint a new and startling picture of the role of teleology and agency in all of life. And joining these two formerly discarded ideas is a surprising finding—the presence of cognition in the simplest forms of life.

## The Future of Biology

We have seen that biologists have historically been strongly wedded to a reductionist viewpoint about life and its basic operations. There has been a common underlying assumption that—just like machines, volcanoes, chemical reactions, and storm systems—living creatures act according to and are influenced by only the well-known laws and forces of physics and chemistry. Some philosophers claim that even humans consist of nothing but electrical impulses, hormonal chemistry, and various physiological and biochemical processes that fool us into thinking we are actually feeling, thinking, conscious free agents with willpower and purpose.

Of course, we are always in the middle of history and not at its end point. We don't know how things will turn out, but I believe we might be on the verge of major fundamental breakthroughs in biology—breakthroughs that will require boldness and a willingness to shed the ideological constraints of the past. Once we accept that biology is not a science based on outdated historical ties to reductionism, we can make even more rapid progress in areas like abiogenesis, nongenetic mechanisms of control, and perhaps even consciousness.

In the rest of this book, I will show you how new, alternative ways of thinking about biology are leading to the ideological shift needed for such progress. And, most important for Christians, it is my strong conviction that with this progress, and with establishing that the science of life is much more than something dependent solely on the laws of physics and chemistry, we will again find pointers to the creator God.

I should clarify that I doubt that science, including the science of biology, will ever "prove" the existence of God. Science is simply not in the business of providing proof, which belongs to logic and mathematics only. But it can provide *evidence*, and we already have considerable evidence for the role of a divine hand in the creation and sustenance of our physical world. I believe that new discoveries in biology are poised to provide even more evidence.

But before we examine those discoveries, let's first look at what many Christians immediately think about when they hear biology and God in the same sentence, a subject that has caused a serious rift within the church—the biological theory of evolution by natural selection.

## Evolution: The Christian Controversy

You will find no denial of evolution by natural selection on these pages. Natural selection is indeed a powerful force and a unique property of life not found in chemistry or physics, and that is one reason biology is a unique field of science. But natural selection is an odd scientific force because it is based on a tautology. In the next chapter, I discuss this in detail and show why there is no mathematical law and no usable definition for fitness, the most basic component of natural selection.

In arguing for the toppling of evolution from its current position as the most fundamental theory in biology, I do not question its truth, but I do question the relevance and purpose of the ongoing—and unwarranted—controversy over its acceptance. The main result of such intense arguments over a biological theory is animosity between Christians. This unfortunate and unnecessary disunity in the body of Christ serves no purpose and can be remedied by a deeper understanding of the science.

I hope to show that all the angst and bitterness on both sides is a tempest in a teapot. I believe the disagreements about the reality of evolutionary processes have been blown out of proportion by folks more interested in a culture war than in reconciliation and mutual tolerance of minor disagreements. I am convinced that future generations of Christians will look back at all the current sound and fury with puzzlement, and I pray that future will begin now.

Evolution is not an alternative to the creation of life by a divine designer. Instead, I see evolution as part of the tool set God used to create all of life, and even as part of a new scientific argument for the existence of God. Evolution does not point to "pitiless indifference," as Richard Dawkins claims, but to a caring and omnipotent deity whose purposes include a universe of more than stars and planets, more than neutrinos and quarks, more than waves and particles: a universe containing the means to create whales and wallabies, tigers and trees, limes and lions, and you and me.

It has always amused me when some atheists paint Christianity as dogmatic and rigid, requiring everyone to blindly follow the set of rules and beliefs dictated by the church, and then in the next breath sarcastically exclaim that there are thousands of Christian denominations and sects that can't even agree with each other about which one is right. It doesn't seem to dawn on them that these two complaints are contradictory, since there would be no denominations if Christianity were in fact dogmatic and rigid.

Christians have never, from the beginning, been of one mind. As a Protestant, I respect and value my Catholic brothers and sisters, but I disagree that there is only one "true" church. On the contrary, I think diversity of opinions and views is healthy, at least here on this fallen earth, while we have only glimpses of the full truth.

We only need to go back to the origins of the church, as beautifully described in the book of Acts, to see that Christian uniformity was never a thing and that followers of Jesus have always had to wrestle with disagreements and various interpretations of his life, words, and works. Acts 15 (one of my favorite chapters in the Bible) illustrates the model for how to deal with a strong and fundamental disagreement among believers. It describes the early controversy over how closely those who believe in Jesus should follow the law of Moses. Was the new faith a branch of Judaism (requiring observance of all Jewish laws), or was it something new? This was a practical matter for the growth of the church, since if the former were true, it meant that Gentiles coming to faith in Christ would need to follow Jewish dietary laws and adult men would need to be circumcised (presumably not a strong inducement to join). In Luke's account, a group of believers from the Jerusalem church held to this position and stood up in a meeting to say so.

In response, Peter shared testimony of how the Holy Spirit came to the Gentiles as well as to the Jewish disciples, and Paul and Barnabas spoke of their work among the Gentiles. Finally, James, the brother of our Lord, suggested a compromise: Gentiles only need to follow a few of the dietary rules—don't eat food sacrificed to idols or meat with blood in it—and observe sexual purity, and that's it (see Acts 15:13-21). And so it was, and the rest is history.

Of course, that was only the first instance of Christian disunity, and later controversies were not always settled so easily or equitably. In fact, many (such as the Protestant Reformation) involved violent confrontations.

Today we have many denominations that differ in details of how believers should worship and express their belief in Jesus. I came to faith as an adult, so to me, these differences appear to be minor and of little consequence compared to the overwhelming and amazing truth of the existence and resurrection of Jesus Christ and the joy of knowing his plan for my salvation. But I know that my lack of concern—about things like whether people should be baptized as babies or when they're older, or whether the pope is indeed infallible, or if Protestantism is closer to the faith of the early Christians than other traditions—is not shared by all Christians.

Fortunately, only a minority of modern Christians consider such disagreements to be so vital as to render their opponents heretics.

But there is one area of disagreement, especially among American Christians, that seems to produce outright hostility in the pews, and in my view it is not at all worthy of the degree of disunity and enmity it causes. It concerns the scientific theory of evolution by natural selection as the explanation of biological diversity.

This controversy is relatively new, starting around the end of the nineteenth century and becoming intense only around the 1960s. Several good books have been written about this history,[13] so I won't go into details here.

My own view of arguments about evolution among Christians is that they are not worth having—for three reasons. First, one can be a faithful follower of Jesus while accepting or not accepting evolutionary theory, just as one can lead a perfectly fulfilled and admirable human life regardless of one's views on evolution. The second reason is that, as I will discuss in more detail in chapter 3, the theory of evolution is irrelevant to what I believe are deeper scientific issues in biology related to Christian faith. And finally, with some exceptions, the people who argue the loudest either for or against evolution don't seem to know what they are talking about, either because they do not understand what the theory actually says, or they are not up to date on the changes in the evolutionary theory that make a lot of their arguments obsolete.

How split is the American church on this issue? According to a 2019 Pew report, a substantial majority of mainline denominational Protestants and Catholics held that evolution is true (65 percent and 66 percent, respectively). On the other hand, only 38 percent of evangelical Christians believed that evolution is real.[14] However, that latter figure is dubious, since a follow-up survey found that the answers of evangelicals (but not mainline Christians) depended on how the question was asked. With a different phrasing, the figure for accepting evolution rose to 62 percent for evangelicals, indicating some degree of complexity in how these Christians think about the issue.[15]

The findings suggest that altogether, roughly half of American

Christians accept evolutionary theory, and the rest don't. That is a pretty significant split that arguably reflects other social and political divisions in American society, with some important consequences for Christian unity in general.

This division over evolution can sometimes reach the level of bitter acrimony. Some will say that young earth creationists, who deny the truth of evolution, are anti-science and contributing to the decline of the faith among the young and educated. Others will claim that Christians who are evolutionary creationists accept a false theory and are denying the truth of God's clear words in the Bible.

## The Virtue of Moderation

I would like to believe that despite the harsh rhetoric of some voices in the culture war, a great many Christians have already come to the realization that the evolution issue is just not worth fighting about. The following is a direct quote from my online friend Terry Godfrey, a devout Christian with a background of strict biblical adherence and an eloquent advocate for the truth of Christ:

> Evolution is a side issue for me, kind of like the flood. Among my Christian friends, evolution is never even brought up. We just don't think it's that important where our relationship with God and our love and trust in Jesus is concerned. I've always left the door open for God's use of evolution because of the verses in Genesis that say, "Let the WATERS bring forth abundantly the moving creature that hath life." "Let the EARTH bring forth the living creature after his kind." "The Lord God formed man of the DUST of the ground." I tell them that if I was convinced that God used evolution that I'd be in even more awe and wonder at God's power and creativity than I am.[16]

All I can add to that is "Amen." I would like to follow the peacemaking example of James at the council of Jerusalem and pour the sublime water of Christ's love on this fire of controversy before it comes close to burning down the church. I pray that when you have finished reading this chapter, you will join me in a celebration of Christian unity even in the face of nonuniformity.

To begin, here is something that might come as a surprise: pretty much everyone agrees, at least to some extent, that evolution by natural selection happens. The argument is not about the reality of evolution but about the extent to which that reality can be applied to understand life on earth.

Let me explain. Some Christians—young earth creationists (YECs) and adherents of intelligent design—agree that "microevolution" happens and results in changes within species or even in the emergence of new species within "kinds," but they think new "kinds" cannot arise this way. The term *microevolution* is not used much by scientists outside of intelligent design circles, but it may be defined as the change within a species due to environmental pressures that lead to the selection of individuals and populations that show beneficial variations. While most YECs agree that such variations are caused by mutations, they differ from mainstream scientists by claiming that all mutations are deleterious and lead to a loss of information (which can nevertheless sometimes provide a more adaptive phenotype, like the loss of eyes in animals living in a dark cave without any light).

The current model of the YEC organization Answers in Genesis holds that evolution "ends" above the species level at families or even orders. Nathaniel T. Jeanson and Jason Lisle, two well-known YEC scientists, have said,

> Young-earth creation (YEC) research within the bounds of the scriptural framework has revealed that the created

"kinds" of Genesis 1 appear to be best approximated by the taxonomic rank of family, not species. . . .

In addition, since preliminary studies suggest that a taxonomic rank higher than family may represent the "kind" boundary for some species . . . , the amount of speciation on the YEC timescale may be even higher.[17]

This is why the Creation Museum operated by Answers in Genesis contains exhibits showing what they believe were the "biblical kinds"—for example, a "cat kind," which they present as the common ancestor to modern species like lions, tigers, jaguars, and domestic cats. In fact, the museum uses a phylogenetic tree similar to the one used by mainstream evolutionary biologists to illustrate what they take to be the post-Genesis flood emergence of modern species from their common ancestors by adaptive natural selection. A comparison of phylogenetic trees from Answers in Genesis and from mainstream biology is shown in figure 1.1.

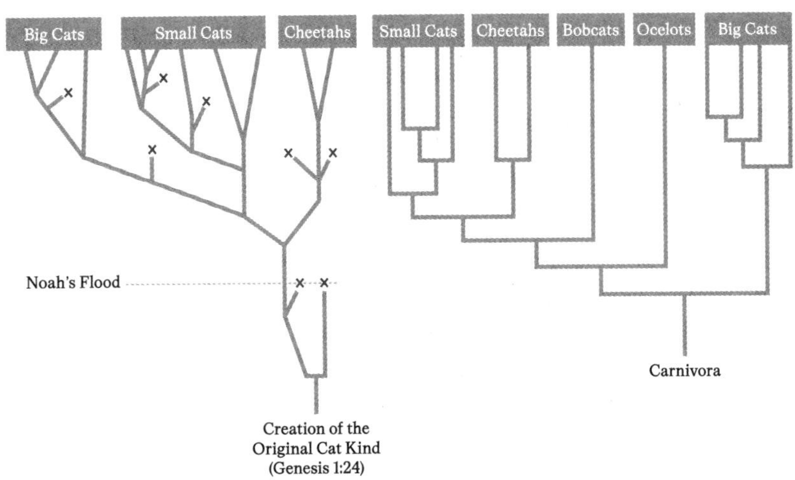


**FIGURE 1.1** *The evolutionary pathway ("phylogenetic tree") of cats is adapted from a young earth creationists organization (Answers in Genesis) on the left,*[18] *and adapted from a mainstream biology paper on the right.*[19]

Mainstream biologists—among them Christians who are evolutionary creationists—hold that "macroevolution" operates according to the same evolutionary principles as "microevolution," and it starts much further back on a phylogenetic tree, beginning with single cells. In other words, one group thinks that evolution by natural selection started operating with single cells, and the other group believes that evolution by natural selection started later, with a set of original created kinds that survived the Flood on the ark. Both groups start with first ancestors whose origin is outside of their respective models. In the case of mainstream biology, the origin of life is outside of evolution and, as of now, lacks a working scientific explanation (more on this in chapters 3 and 5); in the case of Answers in Genesis's kinds, they were of course created directly by God.[20]

I am not trying to minimize the large differences in the scientific views of life between mainstream biologists and those who hold to YEC. But it's worth asking, why do these differences lead to such conflict? Why all the bitterness and strife?

## What Does Evolutionary Theory Actually Say?

Part of the answer is that so many people do not understand what evolution actually is and what it is not. The following from an online opponent of evolution is representative:

> The story goes that there were dark moths and light moths living in England. Industrial pollution or soot started to make the bark of the trees darker. The light-coloured ones stood out more, so the predators could see them easier, and soon there were more dark-coloured moths. They tell you this is EVOLUTION. That the dark ones suited their environment better and natural selection took out the lighter-coloured ones. . . . See. NOTHING HAPPENED.

> The light-coloured ones were removed. That's it. The dark ones already existed. Now there were just more. It's not like the white ones started having darker offspring.[21]

The writer here thinks he's proving evolution wrong, but his comment actually describes the reality of evolution quite well. His comment reflects the common error of thinking that evolution occurs in *individuals* rather than in *populations*. The white moths, in fact, do *not* have darker offspring—they don't have *any* offspring since, unlike the dark moths, they have been eaten. The individual moths don't change; the total population does. The writer says this means evolution does not happen, but what he describes is exactly what evolution is: natural selection acting on preexisting variation.

It's populations that evolve, not individuals. No individual gives birth to a member of a different species. Offspring are rarely radically different from their parents unless there is a strongly deleterious mutation that results in disease or death. In the great majority of cases, they resemble their parents very closely. One would not be able to observe evolution by examining individuals within a population and their offspring.

What is a population? It's a group of individuals of the same species who are able to freely mate with each other. When two groups of the same species are separated by geographical barriers (mountains, deserts, bodies of water, or very long distances), they are two different populations. A good example of this is the famous finches of the Galápagos Islands.[22]

Now we come to how evolution actually works within a population. It's important to note that evolutionary change is generally slow and might not be noticeable in time periods measured in a few lifetimes. If we had two populations living on separate islands that cannot interbreed, and we could observe them over many hundreds of years, we would see that on island 1 the population has undergone some changes (as a group, not limited to one individual), and that on island 2 the population, while the same species, now has a different set of variations. This is usually due to what is called "genetic drift," and it involves random mutations causing minor variations that are mostly neutral—slightly beneficial or slightly deleterious—on each island. Variation already present in the two founding populations can also be amplified over time.

As long as the two populations remain reproductively isolated, the pattern of variations between them will be independent of each other, and after a sufficient amount of time they will begin to appear significantly different from each other. If this difference also involves adaptation to different environments, and the changes become deeper and begin to affect some more basic functions and structures of proteins, there may come a time when the two populations would no longer be able to interbreed. At that point we would have two different species.

I would like to believe that if everyone talking or thinking about evolution could understand these basic realities of the theory, a good deal of the heat of the argument would dissipate. I am aware that this is probably unrealistically optimistic—because, ultimately, the degree of strife is less based on scientific differences or misunderstandings than on much more fundamental disagreements that are cultural in nature.[23]

The religious implications of evolution are often cited as an important source of disagreement. I believe that the modern scientific theory of evolution is not in any conflict with Genesis 1, but most YECs will strongly disagree. When I and my colleagues in science read Genesis, we see a poetic picture of the work of God in creation, an account whose purpose is to express theological truths. We find the text to be a beautifully written declaration of the glory and majesty of God, whose creation of the universe, our planet, and ourselves is the basis of our belief in the omnipotent God we worship. We also see that some of the details necessarily reflect the times in which the text was written—that is, the understanding that ancient Near East people had of the natural world.

In terms of the core teachings of Christianity—trusting Christ's salvific work in the world and loving God and our neighbors—I don't think it matters theologically whether a Christian believes in evolutionary creation or a young earth any more than it matters at what age a person is baptized. And this is why I say that the argument is ultimately not based on scientific or even theological differences but on a cultural divide. Even in the face of strong disagreement, we should unite in focusing on the core of our Christian convictions, free of cultural or political ideologies.

I think that both Genesis and Darwin's theory are correct. God uses both Scripture and science to educate and enlighten us. For me, the truth of evolution is evident, and so is that of Genesis. They do not conflict because their purposes, their explanatory methods, and the questions they answer are not the same but complementary to and harmonious with each other.

As Christians, rather than spending our time and energy arguing with each other about important but secondary matters, let's focus on the words and deeds of our Savior Jesus Christ. That's where we will find our unity, our strength, and our final victory. As Michael Reiss and Michael Ruse, the authors of *The New Biology*,[24] state in a blog post, "The systemic complexity of the new biology points to the inter-connectedness of creation in a way that finds a parallel in the religious vision of the unity of all things in God."[25]

Now that we have tried our best to make peace among warring camps, let's get back to the modern and rapidly evolving science of evolution and see what I mean by everyone getting it wrong, on both sides.

CHAPTER 2

# The Evolution of Evolutionary Theory

SOME SURVEYS (THOUGH NOT ALL) have found that biologists tend to be less religious than physical scientists.[1] If this is true, it's not that physicists see the beauty, majesty, and wonder of the natural world and think about how it all came to be while biologists don't. I think we biologists see at least as much complexity and signs of divine design in the world of living organisms as the physicist sees in the universe and the physical laws that make it work.

But many biologists, while acknowledging the glorious nature of the living world, consider that the theory of evolution by natural selection is sufficient to explain everything about biology, with no role for a divine creator in any aspect of life's origins or development. However, a scientific understanding of the forces, laws, and mechanisms of the natural world does not exclude belief in God as the creator of that world, including all its forces, laws, and mechanisms.

Intelligent atheists have stated that the fine-tuning of the constants in the laws of physics that govern cosmology provides the most convincing scientific evidence for God. Antony Flew, a well-known English philosopher, rejected his atheism based partly on such evidence, as well as the existence of a genetic code.

On the other hand, atheist biologist Richard Dawkins has said that Charles Darwin's theory "made it possible to be an intellectually fulfilled atheist."[2] But even before Dawkins, there was little question that Darwin's theory of evolution helped to spur the rise of atheism, first among educated circles and later in the general public.

But does evolution really align with atheism? Even among Darwin's contemporaries, there were Christians who saw nothing about the new theory that was incompatible with their faith. One of Darwin's strong American supporters was Harvard botanist Asa Gray, a devout Presbyterian who wrote that "God himself is the very last, irreducible causal factor and, hence, the source of all evolutionary change."[3] But the directions evolutionary theory has taken since then, I believe, make it even less compatible with any sort of reductionist, materialist philosophy. I am not alone in that view. Many scientists and laypeople, once they glimpse the majesty of living processes in even the most "insignificant" of creatures, are filled with wonder and awe, and they find the workings of evolutionary biology to be consistent with the work of a creating deity.

So much of what we know about life—about how birds navigate, how whales sing, how insects do the incredible things they do—can be understood as having an evolutionary process as its origin. But to really understand everything about the way life functions, we need to go beyond evolution. In fact, evolutionary theory has already gone beyond the neo-Darwinian idea of evolution.

The time has come to critically examine the widespread assumption that all the wondrous qualities of life are explained, simplified, and rendered ordinary by the natural processes of biological evolution.

I wholeheartedly agree that evolution by variation and natural selection is a powerful and convincing framework to explain the diversity and complexity of living creatures. But there is growing consensus among many biologists that evolution by natural selection as currently understood is not a complete theory. We are missing several key components of how it works and what it can and cannot explain.

We can now see that evolution is not at all atheistic—it can even be seen as evidence for the existence of a creator God. I base that admittedly controversial statement on several emerging facts and ideas from the biological science community.

First, there are some deep deficits in Darwinian evolution as a foundational theory of nature. By *foundational theory*, I mean one that not only underlies a whole world of facts and observations, but one that stands independently of any other, more basic theory. This is how we think of gravitation as the explanation for planetary motion, thermodynamics as the basis for energy and heat transfer, and chemical orbital theory as the underlying principle behind chemical reactions, as three examples from physical science. I will describe some of these deficits below.

Second, the neo-Darwinian concept of random mutations producing changes that lead to natural selection by the blind and impersonal characteristics of the environment, while partly correct, has been modified to include several mechanisms that are neither random nor blind.

To give you an idea of the kind of potentially paradigm-shifting research challenging the outdated concepts of random, purposeless, and blind evolutionary processes, here are a few titles of recent papers from the mainstream scientific literature:

- "How Can Evolution Learn?"[4]
- "Organisms as Agents and Contributors in Ecology, Development, and Evolution"[5]
- "Evolution as Cognition-Based Information Management"[6]

- "Cellular Sentience as the Primary Source of Biological Order and Evolution"[7]
- "Biological Evolution Requires an Emergent, Self-Organizing Principle"[8]
- "Beyond the Modern Synthesis: A Framework for a More Inclusive Biological Synthesis"[9]
- "The Role of Quantum Mechanics in Cognition-Based Evolution"[10]

Third, and finally, as useful as the theory of evolution might be for explaining the diversity of life, it is not, as is often claimed or implied by many biologists (and even more by atheists), the fundamental guiding principle of biology. Scientists agree, for example, that evolution is of no use in tackling the enormous question of the origin of life.

This is why I will argue in the next chapter that the foundational principle of life is not evolution by natural selection but the mechanism of inheritance by cellular self-replication. Evolution follows directly and automatically from accurate self-replication. That principle of inheritance, both necessary and sufficient for the existence of evolution, is a strong pointer to a creative designer of life.

## Current Deficits in Evolutionary Theory

Evolution is not at all mysterious. The scientific definition of evolution is a change in allele frequency in a population over time. An allele is the precise sequence of a specific gene, which can vary slightly among individuals in a population. It is the existence of different alleles that explains why you don't look like me. The only people who have the exact same alleles for all genes are identical twins.

This of course assumes the existence of genes and self-replicating molecules of DNA or RNA as we know them today, but in less technical terms, it simply means that there is variation among the individuals of all populations. Humans have always been aware of this variation and have used selective breeding for millennia to produce sweeter grapes, bigger corn stalks, smarter and more amiable dogs, cattle that give more milk, and so on. We practice "artificial evolution" when we breed the fastest male racehorse with the fastest female racehorse.

The key step in that process is *selection*. Humans consciously and deliberately select the traits (the phenotype) of the animals or plants that we want and breed individuals with those traits to get more of the same or even better. Notice the key process here: *inheritance*. It's another aspect of biology that we have long known about, but back in Darwin's day—and even until 1953—we did not understand how inheritance works scientifically. Now we do understand it in great detail.

Darwin's insight was that in the natural world, things work the same way as they do when humans select favorable traits in order to "improve" a species. However, in nature, there isn't a conscious agent making the selections. Darwin proposed that there is an automatic process of selection, based on relative fitness to the environment.

The word *fitness* has no precise scientific definition in biology—it simply means having the best chance of surviving and successfully reproducing in a particular environment. Prediction is the hallmark of a strong scientific theory, but it is unfortunately impossible to predict *a priori* what phenotypic trait will render a particular population living in a particular environment "fit." Being large, having hair, or even being able to see well could lead to either an increase or a decrease in fitness, depending on the complex details of the local environment. The only way we can quantify fitness in any population is by comparative measurement of survival rates and reproductive success. So when we talk about natural selection and fitness, we're simply saying that the most likely individuals to survive and reproduce in a population are those most likely to survive and reproduce.

This means that our description of the mechanism of evolution by natural selection is ultimately circular. It is a tautology—something that is necessarily true since it follows from itself, and therefore isn't very surprising. For example, if I said good drivers have fewer accidents, and I also said the definition of a good driver is someone who has fewer accidents, I would be making a statement that is undeniably true, but by itself, it tells us very little.

This doesn't mean that evolutionary theory is wrong (as I said, tautologies are necessarily true). But it does mean that while it is well-established, and while masses of experimental data from genetics and molecular biology have confirmed it, we have no good way to frame the theory in rigorous definitional or mathematical language.

There are several useful equations related to population genetics and the role of selection in allele frequencies, but nothing that can rigorously describe the concept of fitness, which remains the center of evolutionary theory. In fact, to date there is no useful mathematical law that fully encompasses the meaning of Darwinian evolution. Ronald Fisher's famous *The Genetical Theory of Natural Selection*, originally published in 1930,[11] describes the general idea of the statistical nature of genetic variation and posits that fitness must always increase. That is useful but not at all comprehensive.

Does this matter? Does a theory need to have a mathematical expression in order to be true? Not at all. The mass of data supporting evolution and the absence of any real data opposed to it are sufficient. Still, it does matter for other reasons—not for establishing the truth of the basic theory but for providing a framework that can be used for testing new ideas and building consensus, among both scientists and laypeople.

For example, once we knew that the laws of thermodynamics and electromagnetism are true and easily described by sets of equations, we could test new ideas to see if they were consistent with or violated these laws. In biology, in the absence of such rigorous laws, the door is easily opened for misinterpretations, misunderstandings, and even distortions of the basic theory.

Mathematical laws are also useful for demolishing the pseudoscience that seems to accompany scientific ideas that are not easily contained in straightforward equations. The lack of such laws in biology has indeed led to a number of bad ideas. Examples of the extension of Darwinian theory to areas that are *not* supported by either the theory itself or any other scientific underpinning include social Darwinism, some forms of simplistic evolutionary psychology, and scientistic theorizing about human morality, the origin of life, or the nature of consciousness.

It is also common for evolution to be used as an analogy or metaphor when talking about nonbiological entities that undergo some kind of selection. Richard Dawkins invented the concept of the meme, the mental equivalent of a gene.[12] Memes—human ideas—compete and reproduce, follow the rules of natural selection, and spread widely or go extinct. In this model, religious beliefs are considered a meme, as are prejudice, fashion statements, popular tastes in food and hairstyles, philosophical and political ideas, and so on. This idea about how ideas spread is quite charming and makes considerable sense, but it isn't biological science.

Technology also seems to follow evolutionary rules. The fittest (meaning most useful, cheapest, or otherwise most popular) new devices survive and edge out those that do not compete as well. One can talk about the evolution of everything from pianos to telephones to calculating devices.

There is clearly something about change that is universal and applies to living creatures as well as human artifacts, thoughts, musical styles, and so on. But let's be very clear: these simple observations about the nature of change and about how so many competing entities survive and prosper (or don't) are *not* what evolutionary theory is about. The theory of natural selection in biology does not apply to memes, disc players, or societies any more than the attraction felt by two lovers is explained by gravitational theory.

Natural selection, as outlined by Darwin and as explained by our current understanding of molecular genetics, involves very specific biological parameters that are not found in these analogous systems. Those parameters, which are uniquely parts of biological systems, include reproduction of the informational DNA molecule, along with a tightly controlled production of phenotype variation, and a method of change that depends on both random and nonrandom mutational events (see below). This is followed by strong interaction with the total environment.

Electronic devices do not reproduce themselves, so the target of selection is not the device but the people who decide if the device is worth manufacturing. In other words, there are some general laws of change, but biological evolution is a very specific and special case of change. Many things evolve (even the theory of evolution itself!), but only in biology does it occur the way it is described in the theory of evolution by natural selection.

## The Evolving Theory of Evolution

In recent decades, a number of new theories have been adding to (though not supplanting) the standard neo-Darwinian paradigm. They are often collectively referred to as the "extended evolutionary synthesis," or the "third way." Unfortunately, the general public is not aware of most of these exciting new findings—something I hope to correct with this book.

A good illustration of the changing tides in biological thought is a recent book titled *Evolution "On Purpose": Teleonomy in Living Systems*. Teleonomy is a form of teleology (purpose-directedness) that is related to a type of program, like the genetic code. Published in 2023 by the MIT Press, the eighteen essays by over twenty contributors cover subjects such as cognition and purpose in evolution, niche construction, and agency in evolution, as well as goal-directedness in biology, emergence, meaning, and other topics I discuss in chapter 4. The authors are well-established scientists and not, at least to my knowledge, active in the science-and-Christian-faith field, and most are probably not theists. The following is a quote from the introduction to the book:

> Perhaps most significant theoretically, there has been a growing recognition that the "purposeful" (teleonomic) behavior of living organisms themselves has had a major influence in shaping the course of evolution over time. Some contemporary theorists have adopted the concept of "agency" to characterize this defining biological characteristic.[13]

The general theory of evolution by natural selection has great explanatory power when it comes to understanding how the vast panoply of complexity in life came to be. But the theory itself has been evolving—continuously—sometimes overturning widely accepted notions and taking new pathways that were initially rejected by mainstream scientists until their legitimacy was firmly established by data and logic. As is the way in all the sciences.

Updates of evolutionary theory are nothing new and have occurred since the start of the twentieth century, when the neo-Darwinian modern synthesis was first accepted by most biologists.[14] Neo-Darwinism emphasized gene-centrism, gradualism (steady, slow, stepwise change), adaptation, and the randomness of mutations.

For the rest of this chapter, I will describe some of the ways in which the theory of evolution is itself evolving in order to accommodate new discoveries in biology. These new discoveries complicate the idea that Darwin's theory of evolution makes it "possible to be an intellectually fulfilled atheist,"[15] as they seem to point to a designer beyond random chance.

### *Punctuated Equilibrium and Neutral Drift*

The ideas of Ernst Mayr and Stephen Jay Gould on punctuated equilibrium[16] were at first roundly rejected by strict neo-Darwinians. After being debated for decades, there is now a body of solid evidence that provides strong molecular mechanisms for the observation that evolution does not always proceed by a continuous slow process of change over long periods of time, as Darwin originally insisted. Instead, the fossil record shows long periods (on the scale of hundreds of millions of years) of very little change, punctuated by brief (in geological time), dramatic moments of explosions of new forms. The Cambrian explosion is the best known of these, but there are many other examples.

The neutral theory of evolution, also known as neutral drift, was first proposed by Japanese biologist Motoo Kimura.[17] Originally highly controversial, it has now become part of mainstream evolutionary theory. It holds that much or most of evolutionary change is not related to selection of better adapted alleles but (especially in small populations) is due to entirely random processes in which a neutral (neither harmful nor beneficial) mutation becomes fixed in a population by chance. While this theory elevates the role of chance in evolution even above the neo-Darwinian view, it also casts some doubt on the idea that positive adaptive mutations are the whole evolutionary story.

### *Epigenetics*

In the nineteenth century, before the Mendelian-genetic model of inheritance was worked out, two similar theories of evolution were proposed. The first one, in 1809, was by French naturalist Jean-Baptiste Lamarck.[18] The second one, of course, was by Charles Darwin, fifty years later. Both men proposed that species can change over time and that such changes can even produce new species, and both saw the role of the environment as critical to this process.

Lamarck's theory was that when an individual organism responded to environmental challenges with increased use or disuse of an organ or process, this not only changed the characteristics of the individual, but these changes could be inherited, resulting in evolution of the population. For example, if the loss of a food source forced an animal population to change their diet, then whatever characteristics they developed to help the new diet work would be inherited by their offspring.

Darwin's theory, on the other hand, saw evolution proceeding in two major steps. First there were heritable changes, and then the environmental conditions selected which organisms would survive and reproduce. Darwin's theory was eventually confirmed, and Lamarckian ideas—the inheritance of acquired characteristics—were rejected both on theoretical and observational grounds.

With the triumph of Darwinism, Lamarckian inheritance continued to be dismissed and even ridiculed until the phenomenon of epigenetics was discovered several decades ago. Studies found that in some circumstances, certain parts of the DNA molecule can be modified in a way that changes the expression of a gene.[19] At first, these findings were not considered terribly dramatic, because it was thought that such effects could not possibly be inherited and therefore could have no impact on basic Darwinian theory. But then it became clear that sometimes these epigenetic changes caused by exposures or environmental influences *can* be inherited, even after the exposure is over.[20] At first it seemed that this only happened for a few generations. But it now appears likely that some epigenetic effects are long-lasting, perhaps permanent, just like a mutation in the DNA sequence.

The crucial difference is that mutations are very rare, accumulate slowly, and are random, while epigenetic effects can be very rapid, are much more frequent, and are targeted to specific genes. The implications of long-term inheritance of epigenetic alteration of genomes are yet to be fully understood and are the subject of intensive ongoing research.

*Convergence*

While some traditional Darwinians insist that purely chance events are the dominant source of variation among individuals, there are new findings from evolutionary biologists suggesting that this might not be entirely true.

Simon Conway Morris's book *Life's Solution: Inevitable Humans in a Lonely Universe* describes myriad examples of evolutionary convergence—how the same complex features, from wings and eyes to social behavior, have evolved many times in different biological lineages.[21] The fact that such similar innovations arose separately in entirely different groups of organisms, often separated by millions of years of evolutionary history, suggests that there could be laws of nature that govern which structures and functions are possible and which are not. This idea goes back to the concept of structuralism proposed by Darwin's contemporary Richard Owen[22] and is still held by some biologists, including Michael Denton.[23]

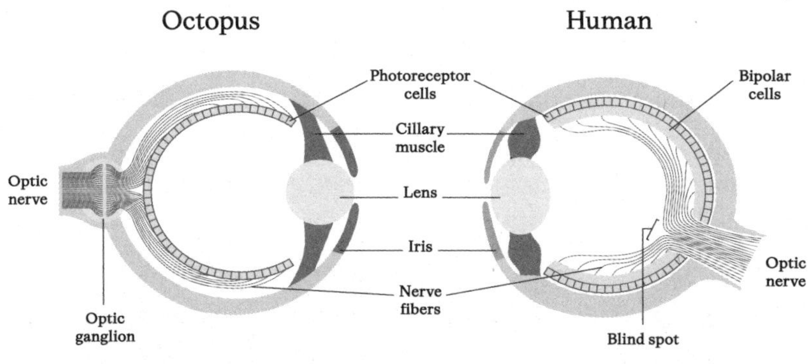


**FIGURE 2.1** *The common structural and functional features of the octopus and human eye are remarkable considering that the two species are separated by hundreds of millions of years of evolutionary history. These features are not therefore explained by any evolutionary pathways, but must result from other, unknown laws of biology.*

While it certainly makes sense that the findings of common directions in evolutionary history point to some underlying laws that determine those directions, no such laws have been discovered as yet. But it seems clear that the idea that evolution is totally blind, that "anything goes" and turning back the clock (as Stephen Jay Gould used to say) would result in an entirely different and entirely random outcome,[24] is false.

### *Robustness and Innovation*

Life faces two contrasting challenges. It needs to have both sufficient flexibility to survive unexpected changes in its environment and sufficient robustness for systems to keep functioning despite perturbations.

Robustness involves a level of stability in the complex mechanisms of metabolism, energy conversion, and gene regulation in the biochemistry of cells and tissues. Without this kind of stability, evolution would not work. Robustness maintains a homeostasis in the face of external change by allowing for duplication and overlap of critical functions so that biological processes are protected from damage to any one part. As a result of robustness, once a novel, improved phenotype is produced, it is stable and resistant to change unless it is replaced by an even more improved phenotype. This is why some features of modern cells have not changed in billions of years.

But most organisms must change in order to survive changing environmental conditions. They do this by undergoing evolutionary innovation, which would appear to be the opposite of robustness. So how does life handle these two contrasting requirements? The answer involves an intriguing solution that leads to a partnership rather than conflict between the two challenges, and both are solved at once.

In order to maintain a high level of robustness, life had to develop and retain many biochemical mechanisms that are rarely if ever used. They are there as backups, in case something goes wrong. For example, a typical bacterium, living happily in your gut, munching on an unlimited supply of sugars in your diet, does not need to be able to metabolize acetone, butane, or glycerol as a source of energy. But it can, because even though the genes that allow for acetone metabolism (for example) are silent most of the time, they can be turned on when needed. The bacterium has all the genes needed for it to live on other sources of energy in case you start dieting and all the sugar goes away.

All living creatures have more genes than they usually need. In fact, we could throw away a large fraction of most organisms' genes, and they would live and function just fine. While this might seem strange, it tells us a lot about how life is able to persevere in the face of unexpected challenges. It also tells us how life is able to innovate.

In his book *Arrival of the Fittest*, biologist Andreas Wagner describes the importance of robustness in our understanding of how life survives, evolves, and creates new functions and features.[25] The theme of the book, and of many of Wagner's scientific publications, is that while robustness and innovation seem to be opposites, they are closely tied to each other. In fact, biological innovation requires robustness, and robustness leads to innovation.

Proteins are robust in terms of their sequences of amino acids and functional activities. Most mutations in proteins have little or no effect on their function, which is their catalytic activity for a specific biochemical reaction. Many mutations can happen without endangering the fitness of the organism. But after many such small changes, a point comes when just one more change can produce a protein with an entirely new function. That is how robustness leads to innovation.

The robustness/innovation connection forms the basis for a new way of thinking about evolution. It provides a mechanism for innovation to occur that doesn't require that every non-deleterious mutation produces an adaptive advantage for the organism. Instead, it provides a way for a series of mutations occurring over time to lead to a major innovative change in function.

### *Nonrandom Mutations*

Of particular concern to many Christians is the idea that Darwinian evolution by natural selection is based, at least partially, on a random process of mutation. The response scientists often give is that the natural-selection aspect of evolution is entirely nonrandom, but not all Christians find this answer convincing enough to view evolution as a nonrandom process. Natural selection can choose which mutations work best (or worst), but it does not choose which mutations will occur in the first place. The theory of evolution has long been understood to be based on purely random mutations that arise, with no connection to what kind of mutation might actually be useful. That is, until fairly recently.

The idea that the source of variation among individuals of a species is random and not in any way directed did not come from Darwin. In *On the Origin of Species*, he states, "I have hitherto sometimes spoken as if the variations . . . had been due to chance. This, of course, is a wholly incorrect expression, but it serves to acknowledge plainly our ignorance of the cause of each particular variation."[26] Darwin did not know how variation arose, and, more importantly, the origin of variation and any distinction between chance and purpose in that process had no consequence for his general theory.

The doctrine of randomness was established in 1943, when Salvador Luria and Max Delbrück showed that mutations in bacteria occurred randomly first and were selected for later, when an environmental challenge arose—in other words, that cells do not choose to produce specific beneficial mutations to help them survive.[27] But then, in 1988, a paper by John Cairns and his colleagues showed that bacteria *could* produce beneficial mutations targeted specifically to relieve severe stress.[28]

Cairns's paper took the world of biology by storm. It was a big step away from the "purely random" concept for mutation and unleashed a major controversy. In the following decades, numerous researchers have confirmed this phenomenon of deliberate production of beneficial mutations (now called "stress-directed mutations") and have found a number of molecular mechanisms to account for it.

Microbiologist Susan Rosenberg has been studying nonrandom mutations in bacteria for decades. Her research has uncovered mechanisms related to the way bacteria respond to stress that show a decidedly nonrandom pattern to the timing and degree of benefit of such mutations. In other words, contrary to biological dogma since the 1940s, life has some degree of control (though we don't know how much yet) over what kinds of mutations take place, when, and where.

A 2019 paper authored by Dr. Rosenberg states that "Mutation is still viewed as probabilistic, not deterministic, but we argue that regulated mutagenesis mechanisms greatly increase the probability that the useful mutations will occur at the right time, thus increasing an organism's ability to evolve and, possibly, in the right places."[29]

### *Niche Construction*

The theory of niche construction (NC) departs from traditional neo-Darwinist evolutionary theory, which focuses on the effect of the environment on living creatures but ignores how creatures affect the environment. Niche construction describes the interactions between organisms and their environment in both directions. For example, beavers make dams, corals make islands, the guano of seabirds produces grasslands,[30] and snails affect the quality of soil.[31] There are many other examples of "the process whereby organisms, through their metabolism, their activities, and their choices, modify their own and/or each other's niches."[32] According to NC theorists, the two-way coevolution of organisms and environments must be taken into account in any comprehensive theory of evolution.

Niche construction, unlike many other aspects of evolutionary theory, is nonrandom and directed, distinct from the effects of random environmental variation on natural selection. Modeling the dynamics of NC-related population genetics gives results very different from the random processes of gene/environmental interaction, where cause and effect go in both directions.[33] An example in human evolution is the spread of lactose tolerance in adults. Dairy farming produced cultural change in the human environment that led to a selective advantage for lactose-tolerant alleles.[34]

By stressing nonrandom, purposeful alterations on the environment caused by the actions of organisms (which then can also have selective effects on other organisms), NC supports the importance of agency in biological evolution, and as such it is an important departure from neo-Darwinism.

### *Cellular Control versus Genetic Determinism*

The Human Genome Project was an enormously successful undertaking. Knowledge of the full sequence of human DNA (and that of many other species) continues to provide a cornucopia of important information, both for medical applications and for understanding the basic biological mechanisms of life.

The project has also come up with some fascinating surprises. The number of genes that code for proteins (structural genes) in humans, which had been estimated to be at least 100,000, turned out to be a very modest 20,000, less than the number of genes in rice, salamanders, and many other much simpler organisms.

So we now know that when it comes to genomes, it isn't the size that matters; it's what you do with it. A related finding, also surprising, was that only about 2 percent of the genome is comprised of genes. We have long known that there is true "junk" DNA in our genomes, long, repeating sequences (some of them the remains of viral infections) and other nonfunctional elements. But among the noncoding DNA—those parts of the genome that are not structural genes—there are also a large number of sequences with a function just as important as that of the genes that code for proteins.

These are the regulatory sequences, and what they do is regulate the expression of the structural genes. For example, everybody has a lactase gene, which codes for the enzyme lactase that is responsible for the metabolism of milk sugar (lactose). The purpose of this enzyme, as in all mammals, is to allow babies to digest their mother's milk. Since adult mammals no longer use milk as food, this gene is turned off (not expressed or downregulated) after weaning age, so very little lactase is made, and adult mammals typically cannot metabolize or tolerate milk. The rare exception to this is a fraction (about a third) of the human population.[35] The cause of lactase persistence (and the resulting lactose tolerance) in these people is one of several mutations, but these mutations are not in the lactase gene. They are found in the genes that *regulate* the lactase gene, and in the case of lactase persistence, those regulating genes keep it turned on. That's why many of us (including me) can drink fresh milk and eat cheese without a problem.

It turns out that the regulation of genes is exceedingly complex and occurs in interactive networks. Each gene in these networks can activate one or more other genes and/or suppress the expression of one or more other genes. We can end up with feedback loops where two genes turn each other on, or one turns on another, which turns off the first one. As you might imagine, things can get pretty complicated. As a result, in some cases it is almost impossible to predict with any certainty what the final phenotype will be (which genes in the network are on and which are off), even if you know the starting conditions.[36]

The fact that gene regulation is so important in the functioning of life has shifted a great deal of attention from the structures and sequences of genes to how and when they are expressed. In terms of evolutionary significance, having a silent gene turned on and making proteins is basically equivalent to having a new functional gene appear. The result is the same: a new protein starts doing its job in the cell, and the phenotype (the actual target of natural selection) changes.

There is still a great deal of both experimental and theoretical work that must be done to fully understand the implications of gene regulation for evolution. However, it is safe to say that the importance of gene regulation, like epigenetics and other newer ideas, questions the notion of genetic determinism. The idea that you are nothing but the sum of your genes has become as untenable scientifically as it is philosophically and spiritually.[37]

### *Natural Genetic Engineering*

Microbiologist James Shapiro has developed a model he calls natural genetic engineering (NGE), in which the cell can control the genome as much as the genome controls the cell. The sources of genetic variation he describes do not fit the neo-Darwinian model of slow, progressive changes—they are rapid, dramatic, and involve grand molecular events.[38]

One example of such a revolutionary event is when the entire genome of an aquatic organism was somehow duplicated. This allowed for descendants of that animal to experience an unusually high rate of mutation without the danger of losing important gene function. As a result of such a risk-free high mutation rate, those descendants rapidly (relatively speaking) developed new anatomical structures, like a backbone, and the entire group of vertebrates came into existence in the form of "bony" fish.

## Design

In his 1802 work *Natural Theology*, English clergyman William Paley famously argued that if we were to find a watch out in nature, we would be able to deduce from the complexity of its workings that it must have been designed by an intelligent watchmaker—that it could not have come into existence by chance.[39] Building on this analogy, Richard Dawkins described the evolutionary process as a

"Blind Watchmaker," emphasizing the role of random chance as the central factor in the evolutionary design of life.

But, again, that understanding is changing. Raymond Noble and Denis Noble published a paper with the provocative title "Was the Watchmaker Blind? Or Was She One-Eyed?"[40] suggesting that evolution is not fully dependent on chance, even in the first stage of producing variation by mutations. The authors propose that cells use various mechanisms to "harness stochasticity," or random chance events, in a way that allows for rapidly finding solutions to challenges.

It is virtually impossible to deny that living things appear to be designed, and no one does. So much of it is so obvious, visible to the naked eye, from the wings of a butterfly to the construction of a seed. Design is everywhere. And this is only the tip of the iceberg. When we go deeply into the workings of the simplest cell, we see amazing designs at work in the biochemical machines (described in the appendix) and in every molecule and functional interaction of life at the microscopic level.

There are other aspects of reality that might appear to be designed, like the geology of our planet and the workings of the solar system, but I personally don't see anything outside of life that bears such obvious features that simply cry out "Thank you" to the designer of such complexity, beauty, and functionality.

Of course, here is where consensus stops. Both Richard Dawkins and Daniel Dennett, two of the original four horsemen of the New Atheism, admit that life appears to be designed. Dennett has said, "The biosphere is utterly saturated with design, with purpose, with reasons."[41] However, he is not speaking of God as the designer but of the mindless, automatic engine of evolution.

My own view is that he is correct, to a point. He says that the appearance of design can be explained by the well-known mechanisms of natural selection. Think of natural selection as a filter. Filters appear to "choose" things, based on size or other characteristics, but we know that they select small or large particles based on the size of their holes, not by conscious decision-making. Natural selection functions the same way, allowing fitter organisms to reproduce and eventually take over the population.

But does the filter of natural selection explain all there is to life? Not at all, as we will see in later chapters. One major example is the origin of life. We know how life brings about new life, including new life forms through evolution. But to elaborate Paley's analogy in a modern setting, when we see robots that make machines, we can tell that at some point there was an original machine designed and made by a human being. So the question we can ask is, Who designed the first life and the evolutionary process that designed everything else? We do not have a scientific answer for that question, as I will demonstrate in chapters 5 and 7.

If as Christians we are interested in the role of God in the creation of life, what we should be looking for are new theories or laws in biology that serve as pointers to God as the designer of life.

While I respect my Christian friends who disagree, I believe we need to accept the emerging new scientific consensus on evolution and use what we know to be true to make discoveries that will lead us deeper. As a committed follower of Jesus Christ as the Truth, I know that a deeper knowledge of the truth about any subject leads always to God.

In the next chapter, I will show why evolution is not the most important aspect of life and that the focus of our interest in theoretical biology should be on something completely different and far more fundamental. The uncanny ability of living cells to replicate themselves with a very high degree of accuracy might be said to make it even easier to be an intellectually fulfilled theist.

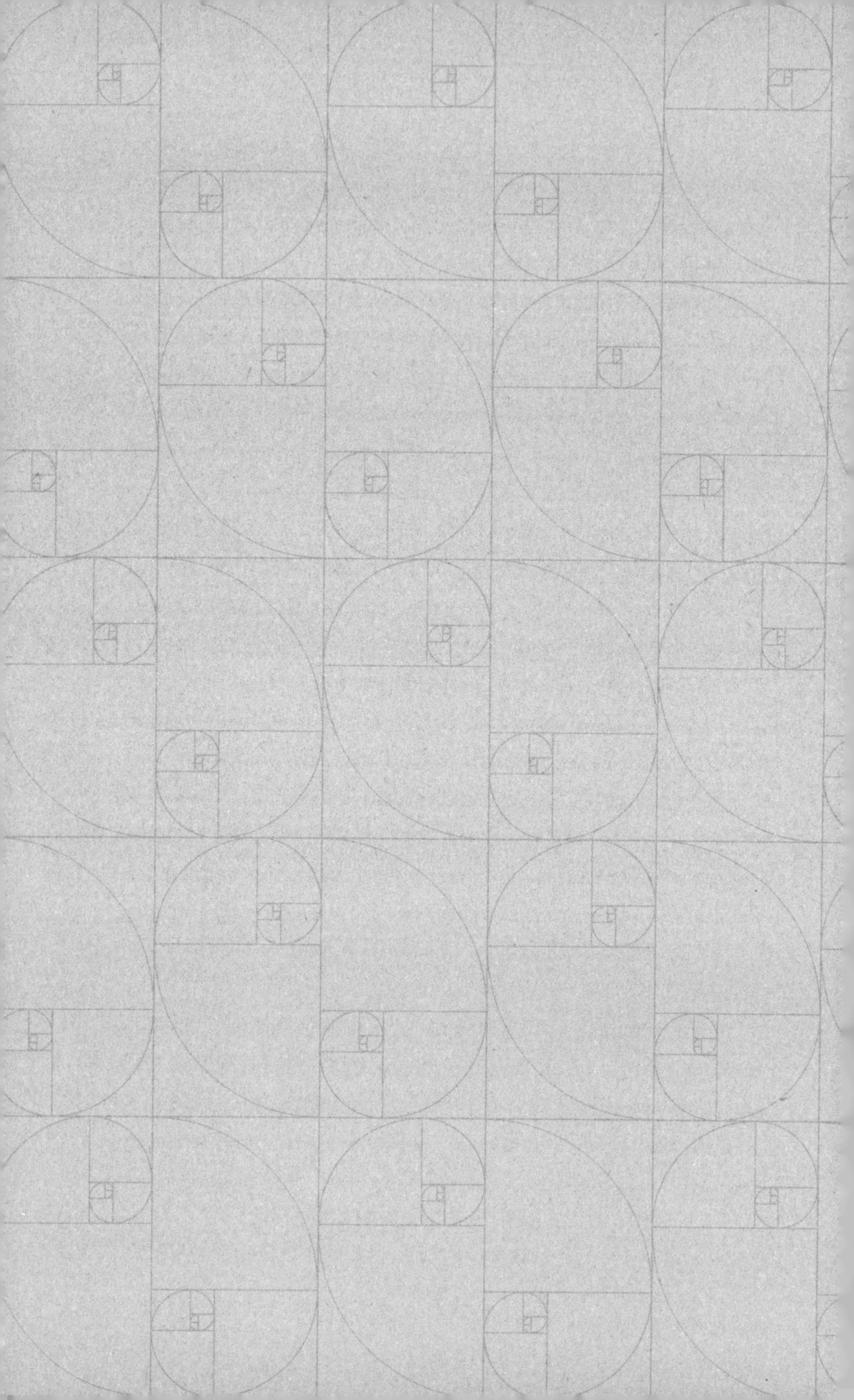

CHAPTER 3

# The Fundamental Principle of Biology: Inheritance

IF YOU WANT TO STUDY BIOLOGY as an undergraduate at most universities, usually you will choose to focus on either molecular and cell biology, which will cover all the stuff I know (like biochemistry and the ways cells function), or evolutionary biology, which covers what most people think about when they hear the word *biology*—things like botany, zoology, animal anatomy, and plant physiology. The naming of the latter shows to what extent evolution is considered not only integral to any study of biology (which, of course, it is) but also the very basis for the entire science of life.

The idea that evolution is the central, fundamental principle of biological science can be found in textbooks, magazine articles, and online discussions on all sorts of topics, from science in general to evidence for atheism or, conversely, for God. I will argue in this chapter that it is time to think about this differently.

This does not mean I think evolution is false. As I stated in the previous chapter, there is no question in my mind that evolution is true, as valid a statement of reality as "things fall down" or "the earth is round." And, again, this also doesn't mean I deny the truth of Scripture. There is nothing about evolution that goes against the authority of the Bible, despite the insistence of some Christian (as well as some Islamic and Jewish) groups to make rejection of evolution an article of faith.

The major point I want to make is that for believers (and even for some scientists), the truth of evolution is not all that important. Evolution says nothing about creation and is simply a natural and *automatic* result of the real bedrock of biological science: inheritance. Inheritance produced by molecular replication in living cells is the true fundamental grounding for life. Evolution follows replication the way falling objects follow gravity, and therefore arguments about evolution, based on either science or theology, have little value or purpose.

## The Strong Principle of Inheritance

Since the dawn of humanity people have observed that things fall down and never up. By the time of the Renaissance in Europe, this well-known phenomenon caught the attention of that small group of philosophers who were interested in finding out how things worked and had the nerve to propose that they might be able to do so on their own, without having to resort to ancient wisdom. These people, later called natural philosophers and then scientists, included quite a few who did many experiments involving dropping things and taking notes. One might say they were working on a theory of "things falling down."

Then one of them, Isaac Newton, came up with a good one: the theory of gravity. In addition to explaining why things fall down, it also accounted for how the planets and the moon moved. He even devised a mathematical law that worked incredibly well to confirm the theory. And as it turned out, the reason things fall down was completely explained by the new theory and the mathematical law.

I believe the origin of species, the universally accepted idea of microevolution, and all aspects of evolutionary biology—including natural selection, individual variation, mutations, and so on—are analogous to the theory of things falling down: they are all explained by a more fundamental principle.

Darwin himself mentioned it in one of the most important and most overlooked sections of *On the Origin of Species*:

> But if variations useful to any organic being do occur, assuredly individuals thus characterised will have the best chance of being preserved in the struggle for life; and from *the strong principle of inheritance* they will tend to produce offspring similarly characterised.[1]

Darwin clearly understood the role that this "strong principle" played in his theory, though neither he nor anyone else at the time knew how inheritance worked, other than simply observing that people and animals inherit many characteristics of their parents and that plants grow according to the nature of their seeds.

We now understand how all living creatures, from bacteria to whales, pass down their characteristics (called the phenotype) to their offspring. We know the molecular biology of the process in extreme detail. And we also know, although almost nobody ever mentions it, that the process of self-replication with very high accuracy, not only of structure but also of function, is found in *all* of life, and *only* in life.

I said that self-replication leads automatically to evolution by natural selection, much as gravity leads automatically to things falling down. We also know that without strong enough gravity acting on them, things do not fall down, as any astronaut will attest. And without accurate enough self-replication, there is no inheritance, and thus no evolution.

Things fall down because there is a fundamental principle in the universe that matter attracts matter. Objects made of atoms have something called mass, and the more mass they have, the stronger is the force of the attraction, which also depends on how close two objects are to each other. Things fall down because the earth is a huge object of immense mass, and therefore its gravitational force pulls down everything nearby.

The same idea is true for accurate self-replication, the fundamental principle of biology. All living forms have the ability to replicate themselves with high (but not perfect) accuracy, and this principle leads automatically to variations, natural selection, and evolution. Once we know that self-replication is the basic mechanism of life, evolutionary change is inevitable and expected.

But there is a mystery here. We know that gravitation is a real force, but we don't know its origin. We also know that cells and other life forms replicate themselves, but we don't know the origin of that phenomenon either. We also cannot invoke evolutionary mechanisms to explain it since (as I will show) the complex system of self-replication had to come first. I believe the mystery of the origin of self-replication will not be solved by any application of the known laws of physics, chemistry, or biology.

## Self-Replication: A Fable

Before we dive into the details of biological inheritance, consider the following fable.

Once upon a time, there was a king who was exceedingly wise and good. His wisdom extended to many areas of philosophy, mathematics, aesthetics, literature, rhetoric, and city planning. The last topic was of great importance to him because most of the people in his kingdom were still living in simple huts scattered all over the countryside, and he wanted to build impressive cities with new, beautiful buildings for them to move into. He was a compassionate king who wanted only the best for his people.

The king was also known for being not entirely practical when it came time to *do* things, and he relied on several advisers, one of whom was called "the mechanic," a man as exceedingly practical as the king was wise.

One day the king came to his council of advisers and told them, "I have a brand-new idea for a city we're going to build. We want people to be happy in their new homes, but we don't know what kind of buildings will make them happiest. Should they be small and cozy or tall and imposing? Should they be heated by burning wood or coal or just well insulated? Maybe we should even put bathrooms inside the houses." That last one got an appreciative chuckle from the others.

"So here is my idea," he continued. "We will build many houses of all different kinds—different shapes, different sizes, different materials—and see which ones the people like best. We will then get rid of the others and build more of the most favored ones. And we can keep doing this, building newer designs and keeping the ones people select as the best."

The advisers were overwhelmed by this latest wisdom from their liege. All except the mechanic, that is. He wasted no time in speaking up.

"Have you considered the expense both in labor and materials of building an entire city, then knocking down most of it, and then rebuilding it, and then doing that over and over again?"

The issue of money was serious, as it always has been, and the council quickly fell silent.

But the king only smiled. "Of course, my friend, and that is where my second, even more ingenious idea comes in!" His Majesty did not include modesty among his many majestic attributes. "We will have no problem with the expense of labor, because the buildings will be constructed in a way that allows them to reproduce themselves."

Now even the mechanic was silent. Finally, he spoke one word: "How?"

"Oh," said the king, "I leave that to you." And off he went.

Try as he might, the poor mechanic, along with his team of architects, engineers, builders, masons, and assorted wise men (whom the king called "scientists" due to their vast knowledge on all matters) could not figure out how a house could reproduce itself. They couldn't get even a single brick to make another brick just like it. They were stumped. Finally, one of the scientists said, "There must be a way to do this. After all, our kingdom is full of noisy children, and all the animals reproduce on a regular basis. So it can't be impossible."

The mechanic nodded and then explained the scientist's error. "Sure," he said. "But that's because people and animals are *alive*. And life is different—life is not a science."

The dreamt-of city was never built, and the king eventually retired to a nice cottage in the countryside. The mechanic went on to build the first computer, but that's an entirely different story.

## Life Is Different

As we learn about how the mechanic's dilemma was solved by the earliest life forms, we must keep in mind that he was right: the things that life can do, especially reproduce itself, are not found in any other kind of matter. The king in the story, while full of brilliant ideas, was not divine, which is why he could not answer the mechanic's question—"how?"

Living cells replicate themselves. Replication is, in fact, an alternative to immortality. Once life can self-replicate, the death of any individual is no longer a major problem, and life as a whole can continue to exist without the strong likelihood of extinction.

We take such replication of life for granted. Animals give birth to babies that are much like their parents; plants spread seeds containing the germs of new plants; and single cells split in two, with each "daughter" cell almost identical to the parent and each other. But if we think about replication as physicists would, this universal phenomenon of biology is quite strange. Nothing else in the physical world does this. A storm might split into two separate storms, but the probability that they will continue to resemble each other as time passes is negligible.

Fire spreads but does not replicate itself. Planets don't split into two identical objects, and neither do stars, stones, staircases, or stereos. But streptococci, stegosauruses, strawberries, and stevedores can replicate themselves with very high fidelity.

## Abiotic Self-Replication

When I make the claim that nothing other than life can replicate itself in casual conversation or on social media, people often push back. I am told that many things in the natural and manmade world can self-replicate. This is not true—not in the sense I am talking about—and it's worth exploring this question in some detail because if I am wrong, then my entire thesis is out the window. So let's take a closer look at alleged abiotic (nonliving) self-replication.

The examples usually given include crystals, self-replicating robots, and even a LEGO machine that can make a copy of itself. But as often happens with origin-of-life research, various fields of engineering and technology are easily distorted when ideas or preliminary experiments are translated for the general public.

The idea that crystals can self-replicate is quite old and connected to a theory that crystals of clay might have been the key ingredient in the origin of life. There is no question that crystals "grow." When a small crystal of defined chemical structure like salt develops in an environment full of the chemicals that make up the crystal (sodium and chlorine), with time the atoms attach themselves to the existing structure in precisely the same arrangement, and the crystal gets larger. The clay theory held that as the crystal grew, parts would break off and form into new crystals, each identical in composition and structure to the original. It was also noted that growing crystals almost always have small imperfections, and the idea that these imperfections might also be replicated after the original crystal broke led to a hypothesis that the division of the main crystal into two is analogous to cell division, with each new generation of crystals faithfully copying the irregularities of its parent. Now add DNA or protein to the clay, and presto! The beginning of life!

It was a good idea, but it was never tested until 2007, when a research group did some thorough and convincing experiments on how crystals divide.[2] The paper is thorough, and the details are technical, but the conclusion is clear. When crystals break in two, they form new imperfections, and these are nonsymmetrical, so the offspring are not at all similar. And with continued growth and division, more and more differences appear in the next generation. There is, as the paper puts it, "more mutation than inheritance." So yes, crystals can replicate, but with very poor accuracy, and the process is not at all similar to the high-accuracy self-replication (HASR) of living cells.

If we compare cells to crystals, we see that the degree of accuracy in biological self-replication is very important. If cells worked like crystals and each generation accumulated more errors than the last, evolution would be impossible. Without a minimum level of accuracy, any beneficial or positive changes in a parent cell would not be evenly passed on to offspring, and eventually so many errors would accumulate that life would end. Crystals have no mechanism to correct the errors and ensure accuracy.

The idea of self-replicating machines and factories has also been around for a long time, and NASA has been very interested in this.

Many plans and theoretical projections on how to construct a self-sustaining and even self-replicating factory in space (for example) might be feasible. But such a system has not actually been made.[3]

There are some ingenious examples of nanobots that do self-replicate in unusual ways that are never seen in life itself. While these systems are fascinating and worth studying, they rely on living cells, so they don't count as abiotic self-replication.[4]

The example of the self-replicating LEGO machine is a good lesson in the problem of hype in this area. An ingenious inventor has made (and demonstrated on YouTube, of course) the workings of a machine made from LEGOs that "builds copies of itself."[5] The problem is that those copies are only small replicas of the outer shell of the machine and have no actual machine inside them, and thus cannot make further copies of themselves. The result is clever and entertaining but not a self-replicating machine at all.

Similarly, a 3D printer can make many different things, such as sculptures, parts for cars or airplanes, and even replacement hips and dental material. The question is, can a 3D printer make an entirely new 3D printer? The answer is sort of. A 3D printer can make many of the components of a 3D printer, but it cannot put them together. That would require another machine that could assemble the parts correctly.

In other words, the most technologically advanced machinery we have at this time cannot self-replicate in a functionally accurate way, while life replicates not only structures but functions as well.

But even if one day humans manage to build something that can copy itself, it will still be questionable whether such a thing can be considered truly abiotic since the intervention by a living human is initially required. The same thing is true for self-replicating computer programs (which are not only abiotic but also not physical).

So, we come back to the unanswered question the mechanic asked the king: How do we build something that self-replicates? If we continue with the idea of a house and include everything in the house—the furniture, books, clothing, paintings on the wall, everything in all the closets—we have a pretty good analogy to what all living cells do routinely. And a cell is far more complex and contains a lot more items than any house. A single cell has thousands of different protein and RNA molecules, lipids, carbohydrates, and other chemicals that include metabolites, precursors, breakdown products, and cellular structures.

In order for a house to replicate itself, it would need to contain a large number of machines that would be able to reproduce the plumbing, electrical, heating, and cooling systems; machines that could make clothing, books, and paintings; and machines that could find all the materials needed and bring them inside. That sounds pretty difficult, but the hardest part is that such a self-replicating house would also need machines that could *make* all those machines or find machines that could make themselves. And in order to do any of this, we would need a plan—blueprints, or maybe even a conscious architect or engineer, or at least a really good mechanic.

We know that life does exactly all of that. But how?

We know the answer. It took a couple of decades to find it all out, using the new science called molecular biology. One of the key parts of the answer is that every cell contains not just chemicals, but actual machinery—*biochemical* machinery.

## Biochemical Machines

Machines made by humans are composed of many materials: metals, wood, plastic, leather straps, ropes—whatever works best. Biological machines are made of unique chemicals found only in life. Proteins, which are important in all aspects of life, are probably the most common materials, but RNA, lipids, and other polymers are also used. For example, membranes (yes, membranes are machines) are mixtures of proteins and lipids containing thousands of smaller machines (receptors, portals, switches) embedded within them. The hundreds of mitochondria that produce all the chemical energy needed for life in cells are machines containing many smaller machines (see part 2 of the appendix on pages 199–205).

If we look at a cell with a high-powered microscope, we can see the larger machines as physical objects composed of hundreds of molecules. But most of these machines consist of many other machines, some of which are single molecules.

The argument I'm making in this chapter—that evolution by natural selection is the natural consequence of high-accuracy self-replication, and therefore, HASR could not be the result of evolution—will make more sense once you see some of the complexity of the biochemical machines of the cell. All of this complex machinery had to be in place *before* evolution by natural selection could occur.

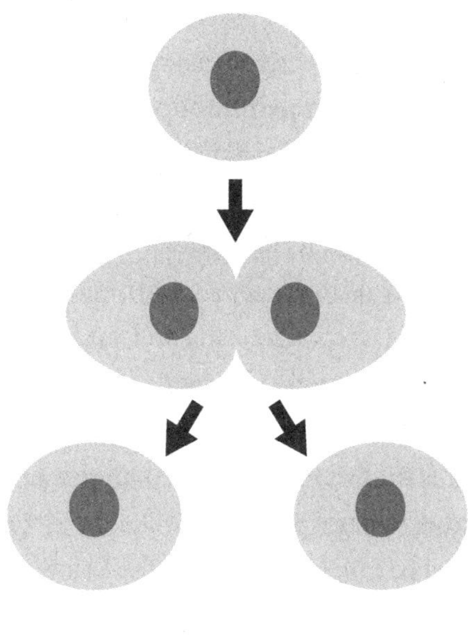

**FIGURE 3.1** *Cells replicate by making a copy of all their biochemical machinery and structures (like the nucleus with its DNA shown here) and then splitting in two to produce two new cells that are over 99 percent identical to the original.*

### *Proteins*

Among the simplest machines in cells are proteins. You have heard of proteins, but you might not know that proteins are the most essential chemicals in all cells. Proteins do everything in life. They are the mechanics and the cooks, the teachers, the cops, and the regulators. How can chemical molecules, even long molecules like proteins, do so many things? The answer is catalysis. Most proteins are enzymes, and enzymes are catalysts. A catalyst is a substance that speeds up change, such as a chemical reaction.

In order to exist, life as we know it needs thousands of chemical reactions to occur at great speed. Biochemical reactions are hundreds or thousands of times faster than chemical reactions in the ocean, the atmosphere, or in the ground because they are all driven by enzyme catalysts.

Every chemical reaction in all life forms is catalyzed by a specific enzyme. Each enzyme does one job, and it does it astonishingly well. In many cases, enzyme-catalyzed chemical reactions are part of a series of such reactions (with each step catalyzed by a different enzyme) that results in the large-scale production of chemicals or molecular structures needed to make a cell and bring it to life.

One of the secrets of cellular self-replication is that once a cell can make copies of all the protein enzymes it needs, those proteins act as machine-making machines. And it turns out that life is great at replicating all the proteins in a cell.

What this means is it isn't necessary for any of the chemicals or structures of a cell to replicate themselves—they are made from scratch by the replicated proteins.

But how do all those proteins get replicated with enough precision that they work as well as the originals? And what about that all-important blueprint, written by and read by the mechanics or builders? How is this biochemical replication planned, controlled, and checked for errors?

These questions are discussed in detail in the appendix, but we'll get a glimpse of the complex processes below. First, let's take a general look at the biological equivalent of the mechanic's instruction manual or the cook's recipe. This is the repository of all the information needed to make the right proteins at the right time for the right job—the collection of genes, called the genome, made of DNA.

### *DNA: The Informational Storage Molecule*

Scientists have long been aware of the idea of a "gene" but had not been able to identify its chemical form. In 1953 James Watson and Francis Crick discovered the detailed chemical structure of DNA,[6] which immediately pointed to how it might function as an information repository. As described in the appendix, that observation unlocked the mystery of how the information in cells is written, passed down, and translated into the detailed instructions for making every protein in the cell.

### *Biological Information*

There are two major definitions of information that are not compatible with each other. The definition of information as used in physics is all about the measure of the uncertainty in a system that can exist in a number of different states. This is often called "Shannon information."[7]

In the context of normal common usage, the word "information" refers to something quite different: the usefulness or meaning of a particular physical state or event. In biology, when we discuss information, we generally mean the common definition that includes meaning, not the one from physics. Some writers like Stephen Meyer use phrases like "specified information" or "functional information" to clarify what is meant by information in biology.[8]

While the transmission of information can be subject to loss of accuracy in many physical systems, it isn't in biology. This is due to a complex and highly efficient set of repair-and-editing systems that are constantly checking and fixing errors in the transmission of information as they occur. Information plays a critical role in the existence of life, and the best example of such biological information is the masterpiece of the genetic code.

### *The Genetic Code*

Living organisms are the only entities we know about in nature that use symbolic meaning, or codes. Humans are experts at coding. Words like *tree* are codes for things. The word *tree*, whether spoken or written, has nothing to do with the tall wooden plant with leaves. Codes are messages whose meaning has to be learned in order to be useful. We learn languages in childhood, and then we can get what we need without crying or having a tantrum.

We are not the only organisms using codes. Bees, ants, and in fact all of life—even bacteria—communicate using chemical codes (see the next chapter). All of this symbolic information derives from the first code, the very first meaningful symbolic information in the history of the universe: the genetic code.

A good way to picture how the genetic code works and what it does is to think of an analogy with an auto repair manual or a cookbook. The letters that make up the words "Replace the faulty sensor" or "Add one teaspoon of sugar" are human constructs that when typed in the correct order convey information to anyone who understands the language. Upon reading these words, the mechanic or the cook then goes to the car or the mixing bowl and carries out the task required by following the informational instructions from the book.

But you might ask, "Isn't DNA a chemical? Doesn't it just do what chemicals do? What makes the genetic code a *code*?" I have heard this from many atheists who do not want to accept that life contains a strong indicator of something beyond the rules of chemistry, something that might even imply an unknown Coder, who designed the code, much as we humans design computer programs and other codes.

A code is a set of symbols that can be uniquely mapped onto (or *mean*) a different and independent set of symbols. For Morse code, this would correspond to patterns of dots and dashes mapping onto letters. For the cookbook, the letters are mapped onto words, which have a meaning to the cook.

For the genetic code, the three-base sequences (*codons*) in DNA are mapped onto specific amino acids that make up proteins in the same order as the order of the codons. There is no physical or chemical reason for any set of bases to *mean* any particular amino acid, any more than there is any reason for *tree* to *mean* a tall plant, which is why that plant can be represented by other words in other languages (like *baum* or *albero*). But unlike human language, life has only one language, the universal genetic code.

## The Genetic Code at Work

We have still not approached the question of how the genetic code works to make just the right proteins needed for replicating everything else in the cell. Without getting too technical, let's take a closer look at this entire system (with much more detail in the appendix). One reason is so you will be impressed and astonished at the ingenuity of the design that governs all of life and, if you haven't already, begin to think deeply about this kind of evidence for a divine designer. But also, once you see the complexity of the system that makes evolution by natural selection not only possible but inevitable, you will see that such complexity could not be the result of the process it must precede.

Suppose we were to assign the task of designing a system that can make the right proteins to a being beyond the level of humans, since no human could possibly come up with a working model of how to make such complex molecules using only the rules of chemistry and information transfer. Perhaps we could describe this being as an angel, known for getting the impossible done.

The angel begins to tackle the problem theoretically, reasoning as follows: "What if the sequence of the nucleotide bases in DNA could be translated into the sequence of amino acids?" (Note: Again, it's okay if you don't understand everything here. I want to give you a sense of the process's complexity!) The angel continues, "There are four bases and twenty amino acids, so we would need a three-base series (a codon) to code for each amino acid. Let's copy the sequence of the DNA onto RNA, which can be the information messenger (mRNA). It will carry a sequence of the three-base codons containing the information to make the right protein.

"Meanwhile, we need a translator molecule that can carry a sequence complementary to the codon that will bind to the codon using base-pairing rules (this means there must be specific interactions between two bases by weak hydrogen bonds) on the mRNA. And the translator molecule must also carry the right amino acid that is coded for by that codon. What can do that?

"Something that is a nucleic acid (since it needs to bind to the mRNA) and also a little bit like a protein (since it also has to bind to an amino acid). We will invent 'transfer RNA'—a small RNA with the shape of a protein that has in it a nucleotide sequence that is the exact complement of the codon (we'll call that the *anticodon*). Now, we need to get the right amino acid onto the right tRNA (the one with the proper anticodon for that amino acid).

"Let's use an enzyme! Aminoacyl-tRNA synthetase (aaRS). We'll have a structure (let's call it a ribosome) where the messenger RNA with all the codons and the 60 different tRNAs containing an anticodon that matches a codon for its amino acid and also carrying that specific amino acid can meet up and allow those amino acids to get close enough to each other, in the right order to bind sequentially, forming a polymeric, linear chain of amino acids, or a protein, of just the right sequence, which will then be released into the cell."

That is one smart angel! And that is exactly how the correct proteins for a cell are made, and those proteins then catalyze all the reactions that manufacture all the other components needed to replicate every part of the cell with close-to-perfect accuracy. But here's the million-dollar question: How did a process that complex develop from a collection of chemicals enclosed in some kind of bubble?

One answer that is offered to that question is that the system evolved by natural selection. But is it possible that the machinery required for cellular self-replication could have evolved by natural selection as we know it?

The answer to this question is straightforward and simple—no! Cellular self-replication *requires* the complex protein-synthesis system and all of its components, including a code carrying the required information. That means evolution cannot *precede* this system and that high-accuracy self-replication had to have come *before* the origin of evolution. The protein-synthesis system could not have evolved through a process that requires it to already be in place.

Without a system for protein synthesis and cellular self-replication, there can be no evolution. But once a sequence-specific protein synthesis system exists, even if not perfectly accurate, then cells containing more efficient catalytic enzymes can pass on the ability to make those proteins to daughter cells, and the frequency of the better variants can increase with time, which is the definition of evolution. Assuming that system includes some form of information-carrying gene, there is now a mechanism to ensure that such evolution is possible.

That also means that once self-replication becomes possible, evolution is not only *possible* but happens *automatically*. I call this inheritance first theory (IFT), since it states that accurate inheritance must come before the origin of evolution.

## Inheritance First Theory (IFT)

If high-accuracy self-replication (HASR) exists, then evolutionary change is inevitable, because HASR includes a small but significant amount of variation in each generation, and those variations will result in a differential fitness of individuals. But if HASR is absent, then any beneficial variation in an individual will not be copied sufficiently accurately in future generations, thus removing it from contributing to evolutionary change.

The scientific definition of evolution is a positive change in the frequency of the fittest allele over generations. We can quantify the degree of HASR by using the probability that the fittest allele will be accurately reproduced between generations.

I recently published a peer-reviewed paper in the journal *BioSystems* that demonstrates mathematically that the accurate replication of a living cell is the fundamental feature of life that allows for evolution. In this paper, I show that if a cell divides in two but there is no mechanism to make sure the two cells are very closely similar in all their characteristics to each other and to the parent cell, then evolution is impossible.[9]

But if the degree of accuracy in cellular self-replication is above a particular threshold, and if there is a difference between the fitness of different variants within the cell population, then evolution will automatically take place. This means that fitness variation and accurate inheritance are together both necessary and sufficient for evolution.

The required *threshold* of replication accuracy is related to the difference between the fitness of the variants within the population. The smaller the fitness differences between the variants, the greater the replication accuracy must be in order for evolution to take place. (More details from this paper can be found in the appendix—see pages 181–205.)[10]

The major takeaway is that there must be at least some degree of replication accuracy for evolution to occur, and in most cases the threshold requires at least 50 percent accuracy or higher.

To summarize the import of these findings, I say in my paper, "High-accuracy self-replication is the central issue in finding mechanisms for the origin of evolution and therefore of life as we know it. . . . In the grand scheme of the underlying foundational principle of life, we can conclude that self-replication, rather than evolution, is primary."

A similar assertion was recently published in the *Proceedings of the National Academy of Sciences of the United States of America* by Eugene Koonin and his colleagues. Two quotes from this paper illustrate the views of a leading (nontheist) scientist in the field of origin of life:

> For natural selection to kick off and to sustain long-term evolution, an essential condition is replication of a complex digital information carrier (a DNA or RNA molecule). The replication fidelity must be sufficiently high to provide for the differential replication of emerging mutants and survival of the fittest ones. . . .
>
> In modern organisms, accurate replication is ensured by elaborate molecular machineries that include not only replication and repair enzymes but also, the entire metabolic network of the cell, which supplies energy and building blocks for replication.[11]

Natural selection in life is an inevitable, passive consequence of molecular replication by a sequence-specific protein-synthesis system. This also means that the information carried in the genes, which is a crucial part of the overall inheritance mechanism, is required for evolution by natural selection.

Since the modern protein-synthesis system (which, we saw, is quite complex) could not have evolved by natural selection as we now know it, some other unknown, more basic and fundamental process that does not rely on differential fitness or molecular replication is required. Despite many decades of scientific research, there are no good ideas or theories as to what that might be. We will tackle this issue more in the next chapter.

Meanwhile, let's see what the Bible has to say.

## The Biblical View of Inheritance

> Then God said, "Let the land produce vegetation: seed-bearing plants and trees on the land that bear fruit with seed in it, *according to their various kinds*." And it was so. The land produced vegetation: plants bearing seed *according to their kinds* and trees bearing fruit with seed in it *according to their kinds*.
>
> GENESIS 1:11-12, EMPHASIS ADDED

Note the emphasized words. With minor variations, they are all saying the same thing: "according to their kind." In verse 21, we see two more examples of this language, with both plural and singular forms:

> So God created the great creatures of the sea and every living thing with which the water teems and that moves about in it, *according to their kinds*, and every winged bird *according to its kind*.

And then in verses 24 and 25, the writer goes wild with five repetitions of the phrase:

> And God said, "Let the land produce living creatures *according to their kinds*: the livestock, the creatures that move

> along the ground, and the wild animals, each *according to its kind*." And it was so. God made the wild animals *according to their kinds*, the livestock *according to their kinds*, and all the creatures that move along the ground *according to their kinds*. And God saw that it was good.

While being careful not to project modern science back into Scripture, we nonetheless cannot but note here a strong emphasis on the importance of inheritance. I am not a biblical scholar, and I know that word repetition in the Bible might not always be as significant as it might seem to us—poetic forms and other ancient traditions that I know little about might also be involved. Still, I believe that something important is being emphasized by this phrase. The original Hebrew root מִין [*min*] means "species or kind," and it has been interpreted in various ways,[12] but its meaning in these verses is clear: living things reproduce and have offspring that resemble themselves closely.

There is no argument against evolution to be found here (contrary to what some people think). Rather, evolution *depends* on exactly the same fact of inheritance that ensures that sheep only give birth to sheep, the seeds of grapevines only and always make new grapevines, and monkeys give birth to monkeys and never to humans.

That living things reproduce "according to their kind" was of course well known to the farming and herding people of the ancient Near East. But if the phrase is simply a repeat of common cultural knowledge, why is it stressed so much? I believe this part of Scripture reflects a deeper truth: central to the creation of life—whether the plants in verses 11 and 12, or the animals in verses 24 and 25—is the fundamental principle of inheritance.

It is that mechanism of high-accuracy self-replication that is described and repeated in Genesis 1. "According to its kind" could be thought of as meaning "accurate replication of phenotype." The additional word "various" in verse 11 is simply icing on the cake.

Variation is the other essential ingredient in the evolutionary process, and here it is in Scripture about life's creation. The Bible is of course not a science textbook, and concordism—trying to fit our modern scientific understandings into the biblical text—is usually a mistake. Still, in this case, it feels like God is telling us that he created life in such a way that every living creature reproduces itself accurately. And so it came to pass. And God saw that it was good.

## Summation

In chapter 1, I argued that evolution is not an issue worthy of strife between Christians and in chapter 2 that the basic theory of evolution is in the process of undergoing major changes, which make many of the arguments about its truth irrelevant. In this chapter I have presented an additional reason for us to stop focusing on evolution as an important issue for debate. Evolution, contrary to what many on both the secular and creationist side think, is not the center of biological science but an automatic outgrowth of the truly central feature of life—accurate inheritance, which both science and the Bible support. The theory that inheritance had to come before evolution and that, once in place, led directly to it has biological, theoretical, mathematical, and biblical evidence behind it.

What does this mean for Christians? Evolution is a scientific theory, and thus purely naturalistic, although many Christians, including myself, can see a role for God using evolution as a tool in his creation of life. Where does high-accuracy self-replication fit into our scientific and theological understanding? We can explain the details of how it works, but its origins are shrouded in mystery.

Mysteries by themselves do not always constitute evidence for a divine creator. But sometimes they do. I strongly believe that the origin of HASR, the fundamental basis for all of life, is like the fine-tuning of the physical constants, the origin of the universe, and the nature of quantum reality. It is one of those cases where scientific knowledge, rather than a gap in such knowledge, is nothing less than evidence for God. To summarize the case, here are the facts we know:

1. Life exists and persists thanks to a biochemical process of enormous complexity, *not found anywhere else*, that allows for all living cells to replicate themselves with a high enough accuracy to allow for evolution to occur.
2. This process is not just chemically complex but involves the invention of abstract information in the form of a code, whose origin is not explained by any known law of chemistry or physics.
3. As with all scientific evidence for God, alternative explanations may be possible (which is why there will never be a scientific proof of God), but in the case of inheritance first theory, there are no such alternatives known.

In light of these facts, I propose that we must add the origin of high-accuracy self-replication of cells to our list of strong evidence for the existence of God.

Am I claiming that the first cell that replicated itself (with high-enough accuracy so that evolution could start) was the result of a miracle? As a follower of Jesus Christ, yes, that is what I claim. I do believe that life is a miracle—just as the origin of the universe is.

But as a scientist, I still want to know how God did it. I don't think we can find that answer using the tools and laws of chemistry and physics that we currently have. However, science is no more "finished" today than it was before Einstein and Planck overturned what we then thought was our ultimate knowledge of physics.

I believe we will continue to make progress in finding new biological laws that will shed some light on all these mysteries. We are already beginning to see the glimmer of such laws, and I have no doubt that just as with all our knowledge of God's creation, they will glorify the source of all of nature, the majestic power of our Lord. The next chapter is all about that.

CHAPTER 4

# Agency, Cognition, and Teleology (ACT) in Biology

EVERY BRANCH OF NATURAL SCIENCE is replete with theories and laws. Scientific theories are statements of how things work, and laws are mathematical constructs of such theories. For example, the theory of gravity states that objects attract each other in a way that depends on their sizes (mass) and the distance between them. Newton's law of gravitation has the form of a mathematical equation: if you take the product of the masses of two objects divided by the square of the distance between them and multiply that by the gravitational constant, you can calculate the exact strength of the gravitational force between the two objects (like the earth and the apple about to fall from the tree).

A theory in science is not a guess—it is a well-established framework based on a preponderance of empirical data and facts, as well as the absence of any data that contradicts it. I remember one of my professors in graduate school paraphrasing Thomas Huxley: "A beautiful theory can be entirely destroyed by a single ugly fact."

Chemistry, physics, and geology are full of theories and laws that students and aspiring scientists must learn. Scientists base their own work on these theories, including research that could extend, modify, or even refute them. But biology is strangely lacking in theories and includes very few general laws.

We can ask why biology is so notoriously short on theory. While there is a field of theoretical biology, and even a journal of that name, biologists have traditionally been hesitant to think of their field as needing a theoretical foundation. I have heard colleagues dismiss the entire concept of theory in biology as impossible or even nonscientific. I believe that is a serious problem.

An argument could be made that despite the unquestionably rapid pace of scientific progress in biology, our actual level of understanding of the mechanisms that operate in living cells is less profound, less integrated, and less easily communicated than it should be, given the enormity of the data that has been generated. We are often faced with fragmented fields and subfields of research into an explosive number of subcellular factors, proteins, genes, and pathways. Each is studied in isolation or in association with a very limited subset of the bewildering complexity that is the reality of the cell. So-called models in biology are generally diagrams of mechanistic reaction pathways that tend to resemble road maps, with lots of arrows pointing to and from long lists of gene products. These may help us learn useful things about some diseases, for example, but they do little to reveal any underlying or general rules of the science.

I believe that the reason for the discrepancy between the vast number of facts we know and the shallowness of our understanding in cell biology stems from a paucity of theoretical efforts in the field. In all other fields of science, theoretical and experimental work tend to progress together, each driving the other and leading to some degree of an integrated framework of understanding.

One possible response to a plea for more theory in biology is that since biological systems are far more complex than the systems physicists deal with, biologists are forced to focus on relatively narrow areas and cannot hope to address global questions with comprehensive theories. But the role of theory in science has always been precisely to take large masses of uncoordinated information that alone seem to suggest unfathomable complexity and use them to create an orderly, logical, testable, and ultimately satisfying understanding of nature.

This is what happened when the theory of gravity explained why things fall down. Charles Darwin, Alfred Russel Wallace, and many other biologists ("naturalists") before them collected vast amounts of data about the morphology and comparative anatomy of species. The world of living nature seemed frustratingly convoluted and messy until the elegance of the theory of evolution by natural selection made sense of all that diversity.

Biologists of Darwin's time were excited by his theory because it advanced the science of biology beyond simply describing the structures and functions of organs and tissues. It provided a sense of order, meaning, and purpose to the way life works. Asa Gray, one of Darwin's earliest supporters, said, "We recognize the great service rendered by Darwin to natural science by restoring teleology to it, so that instead of having morphology against teleology, we shall have henceforth morphology married to teleology."[1]

And Darwin himself, his son Francis Darwin, and Thomas Huxley all agreed. They all saw the theory of biological evolution by natural selection as including the idea of a final cause (telos) that had until then been missing in biology. That final cause, by which Aristotle meant the purpose or end-goal of what was caused, could now be seen as the improvement of the chances for success, or fitness, provided by the process of natural selection.

There is some irony here: despite the vast success of evolutionary theory in the twentieth century, the idea of teleology became highly controversial in biology. Since it is absent in physics and chemistry, many biologists deemed it a nonscientific concept and strove to eliminate it from consideration in biology as well. As I will discuss further, I believe that was a mistake, and Darwin and his colleagues were correct about evolution and teleology.

I fully realize that it's one thing to make a case for the need for more theory in biology, and quite another to actually propose and describe any such theory. And I regret to clarify that you will not find any brilliant ideas here that can pull everything about life together in an elegant, testable, and shiny new theory. Unfortunately, I am not that smart. But I do pay attention, and what I will do is present you with some exciting, newly published information coming from smart biologists that could very well end up being used in the construction of the missing theoretical framework for the science of life.

I will focus here on three concepts that were in the past deliberately discarded from use in biological science because they did not fit the prevailing reductionist ideology of the times. But the times have changed, and teleology and agency are back, joined by an unexpected development: the introduction of cognition where it was never thought to be found—in the simplest life forms.

Even on the smallest scale of single cells, living things ACT—they show *agency*, *cognition*, and *teleology*—which challenges the prevailing materialist, reductionist views in scientific circles. Life follows the same laws as nonliving physical matter, but such laws do not fully cover the reality of how life functions. It is a legitimate question to ask where purpose and agency in life come from. And what gave rise to universal cognition?

I like the acronym ACT because it beautifully spells out one of the key things that life does: it *acts*. I will, however, discuss the concepts of agency, cognition, and teleology in a different order, one that is more suited to making my case.

## Teleology

The word *teleology*, from the Greek *telos*, refers to explanations that consider the purpose or end of something, and it comes from the philosophy of Aristotle, who postulated that there exists for all events a "final cause," or the reason or purpose for the event to occur.[2] While this definition is not strictly accurate in a philosophical sense, I will use teleology as a synonym for purpose in what follows.

Science has always treated the concept of purpose as forbidden when it comes to the laws of physics and chemistry—and for good reason. No storm system cares if it dies out or does not give rise to another storm. No asteroid has any reason to collide with a planet, other than following the laws of motion and gravity. No volcano acts on purpose to erupt—the system simply obeys the physical laws of pressure and heat transfer—and no rock cares about being eroded by a river and deposited as sediment somewhere else. But bacteria, oak trees, and dolphins *do* care. The fact that teleology plays no part in the laws of physics or chemistry says nothing about whether teleology exists in biology.

Lions, foxes, and eagles hunt for a reason. Rabbits and deer run away on purpose. And lions, rabbits, and most other animals protect and care for their young. All living creatures—plants and animals—do all the innumerable things they do *for a purpose*: to stay alive and to reproduce. Unlike rocks, storms, volcanoes, and the rest of the nonliving universe, life operates with purpose. It is goal-oriented.

Many scientists and philosophers have historically been skeptical at best about the idea of purpose in biology, and some have claimed that any form of teleology is contrary to the very fabric of science.

However, I think it is safe to say that the trend toward accepting the reality of purpose throughout the biological world, as formulated by Darwin's theory, is now close to becoming mainstream.

To give you an idea of the kind of paradigm-shifting research challenging the deterministic concepts of how biology, even at the smallest scale, works, here are a few titles of recent papers:

- "All Living Cells Are Cognitive"[3]
- "Biomolecular Basis of Cellular Consciousness via Subcellular Nanobrains"[4]
- "Learning in Single Cell Organisms"[5]
- "How Biological Codes Break Causal Chains to Enable Autonomy for Organisms"[6]
- "A Revised Central Dogma for the 21st Century: All Biology Is Cognitive Information Processing"[7]

These peer-reviewed papers, as well as those cited in chapter 2 (and many more like them)—all proclaiming new ways of thinking about biology—were published in mainstream biological journals. These are not fringe ideas.

I would wager that you might be surprised to learn that scientists are now talking about cognitive cells, information management, and the simplest of life forms acting as agents to determine their own evolutionary path. And if you are curious about what a nanobrain is, along with many more exciting innovations in modern biological science, read on.

J. B. S. Haldane, one of the pioneers of biological theory in genetics, is reported to have expressed his frustration at the unwillingness of biologists of his time to admit the reality of teleology with the famous quip, "Teleology is like a mistress to a biologist: he cannot live without her but he's unwilling to be seen with her in public."[8]

Historian and author Timothy Lenoir put it this way: "Teleological thinking has been steadfastly resisted by modern biology. And yet, in nearly every area of research biologists are hard pressed to find language that does not impute purposiveness to living forms."[9]

There was a time when even stating that "turtles come ashore *in order to* lay their eggs" was seen as too indicative of inherent purpose in an animal's behavior—"turtles come ashore *and* lay their eggs" was considered more "scientific" since it conveys no hint of purposeful action.

This is now generally considered silly, and the use of teleological language has been defended by Ernst Mayr, one of the giants of biological theory and philosophy, who wrote, "'A turtle came ashore to lay her eggs.' . . . There is now complete consensus among biologists that the teleological phrasing of such a statement does not imply any conflict with physicochemical causality."[10]

Teleological-sounding language is unavoidable when teaching or talking about biology. Here are just two examples, one from a scientific paper and the other from a textbook (the teleological words are italicized):

> As a consequence, enzymes *facing* this problem *have evolved* a second active site. . . . These constraints have *guided* the evolution of enzymes involved in information transfer to *reach optimum* ratios of accuracy and speed. . . . How do synthetases *deal* with this?[11]

> Several mechanisms *working together ensure* that the tRNA synthetase *links the correct* amino acid to each tRNA.[12]

Even atheist philosopher Daniel Dennett wrote that "the biosphere is utterly saturated with design, with purpose, with reasons."[13]

### *Purpose in Biochemistry*

Goal-oriented processes are built into the deepest levels of molecular and biochemical functioning of all living creatures, as well as the more obvious purposeful behavior of animals. The critical biochemical pathways that reflect the purposefulness of life include replication of the genetic information, reproduction of the organism, and translation of the genetic information into biological characteristics (the linkage between genotype and phenotype), as well as all the metabolic chemical reactions that convert extracellular material into the chemicals and energy needed for life to persist.

Each of these biochemical processes is highly purposeful because none of them occur at random or depend on accidental appearances of forces or material. Rather, they are well controlled and specific in timing and outcome. For example, the protein enzymes of life are not ordinary catalysts—they are fine-tuned to do their jobs with precision. Unlike nonbiological catalysts, enzymes are highly specific, catalyzing only one reaction involving a single set of reactants, and they are extremely efficient, reaching reaction rates hundreds of thousands of times higher than that of uncatalyzed rates. Enzymes participate in complex control and feedback mechanisms, and they include all the required energy conversion mechanisms to drive reactions. Finally, many of them are produced only when needed—another case of unmistakable purposefulness.

### *Evolutionary Purposes*

But isn't evolution ultimately random? As we have seen, no, not completely. Many of the findings of new evolutionary mechanisms—such as natural genetic engineering, epigenetics, nonrandom mutations, cellular control of genomic function, and complex patterns of gene regulation—are consistent with a goal-driven biological mechanism that has yet to be clearly identified or articulated.

And as evolution produces more and more complex organisms, purpose takes new forms, until we arrive at the kind of creature that mulls over the exact words to write in a chapter about teleology in life for the purpose of best expressing his ideas.

We humans are the ultimate proof of biological purpose, and for that we must thank God for his creation of life in ways that allow for creatures who have the will and purpose of serving and worshiping him. If we believe that God created life and that humans are God's image bearers, then we can think of God's will to have living creatures capable and desirous of a relationship with him as the final cause of evolution.

The acceptance of teleology as part of biology has been growing stronger in mainstream science in recent years. As Denis Noble (a nontheist) has said in relation to the operation of the mammalian immune system, "Organisms and populations of organisms do have identifiable and empirically testable goals. . . . The targeting of variation and the preservation of existing functionality ensure that evolution is not entirely blind."[14]

And as described in chapter 2, publication of the book *Evolution "On Purpose": Teleonomy in Living Systems* contains many more discussions on the growing awareness of the central role of teleology in biology. An additional quote from the introduction of that book relates to the subject of this chapter: "All living organisms utilize a form of sentience, and . . . the emergence of eukaryotes and more complex living systems from prokaryotes required the development of a unitary cognitive system."[15] (Prokaryotes are simpler cells, like bacteria, that have no nucleus, complex chromosomes, or energy-producing mitochondria. Eukaryotes are far more complex, and all multicellular organisms have evolved from eukaryotic cells.)

It is quite possible that by the time you are reading this book, there will be a flood of new papers and books on this theme, and word will have spread to the general public that purpose is a real component of biological science.

### *Biology and Knowledge of the Future*

One of the problems with the idea of purpose in biology is related to the long-standing principle that future consequences of any action or change are not detectable by the organisms making or undergoing that change. This of course doesn't apply to humans, who can (at least to some extent) foresee the consequences of their actions, but it was believed to be true for bacteria, invertebrates, and even most mammals. But as we will see, recent research is questioning this principle, and as the paper authored by Susan Rosenberg says, "Assumptions about the constant, gradual, clock-like, and environmentally blind nature of mutation are ready for retirement."[16]

It now appears that the future consequences of biological change can indeed effect such changes, even for the simplest of organisms.

If the processes of biology are restricted to follow only the laws of chemistry and physics, then no hint of teleology is permissible, and turtles indeed only come ashore *and* lay their eggs because of blind, deterministic instincts built into their nervous systems that cause them to follow an automatic program. But, again, it seems we cannot rely solely on the known laws of physics and chemistry to understand biology. I think if we can find a way to incorporate teleological factors into formal evolutionary theory, it will represent a great scientific advance. We need a stronger explanatory framework for the undeniable presence of teleology in biology—along with its sister phenomenon, agency.

## Agency

Determinism is a way to think about events that occur in the world that leaves little or no room for the existence of free will. According to strict determinism, all events happen as automatic results of a chain of preceding events determined by the laws of physics and chemistry and/or the laws of chance. This *sounds* very scientific, and while determinist arguments have existed since antiquity, it is in the modern scientific age that a significant number of people subscribe to them, including many (though not all) atheists. Most people, however, tend to think that while strict determinism might very well be true for the great majority of things that happen in the universe, when it comes to living things, especially humans, there is another kind of causation: the free will of an agent. This would include you picking up the keys to your car in order to go shopping for your church's food pantry, but it could also be your dog bringing you your shoes when she wants to go for a walk, or a mother fox walking off with your dog's toy from the yard to entertain her kits (as recently reported on our neighborhood social media).

Skeptical philosophers correctly state that purpose denotes agency, and if the biological existence of agency is denied, then so must the idea of purpose. But every living thing is an agent, while rocks, volcanoes, and crystals are not. An agent is an entity that acts intentionally, in a goal-oriented way. Rocks don't act; bacteria do. A boulder rolling down a hillside is not acting; it is simply following Newton's laws of motion and gravity. Living things act—or, as contemporary scientists Denis Noble and Raymond Noble put it, "Organisms are agents to the extent that they can interact socially with other organisms to choose particular forms of behavior in response to environmental challenges."[17]

The agent of these processes (as well as of all the other processes required for life, such as energy conversion, biochemical synthesis, and homeostasis) is the living organism itself.

Along with dismissing teleology, some modern atheists deny the reality of agency as well. Sam Harris, Daniel Dennett, and some of their colleagues in the New Atheist movement claim that agency is an illusion, and that if one looks deeper, it simply disappears. It becomes, like purpose or the concept of the divine, just another form of magical thinking that misleads people into believing there is anything beyond the reductionist, material reality of basic physics.

### *Agency and Free Will*

Sam Harris, one of the four original leaders of the New Atheist movement, argued there is scientific evidence for the nonexistence of free will in his book *Free Will*.[18] He points to, among other research, a series of experiments by neuropsychologist Benjamin Libet in the 1980s, which found evidence for unconscious brain activity indicating action a second or so before subjects were aware they made a decision to act.[19] Harris (and many others) have concluded from this that unknown neurochemical stimuli, rather than conscious thoughts, are responsible for our decisions. In this view, our perception of deciding to act comes later, after the action was already started by unconscious neural activity, and agency is therefore a mere illusion.

That certainly sounds "sciency" and has convinced a lot of people, but others point out that these experiments were deeply flawed. Psychologist Steve Taylor, author of *Spiritual Science: Why Science Needs Spirituality to Make Sense of the World*,[20] explains the many problems with the Libet experiments in an article in *Scientific American* titled "How a Flawed Experiment 'Proved' That Free Will Doesn't Exist," which makes the case that "it did no such thing—but the result has become conventional wisdom nevertheless."[21]

Taylor catalogs a number of serious deficiencies in the experiment's design, including the fact that making a decision to push a button and assigning a time on the clock to the moment of making that decision are not the same thing: "There may be a delay between the impulse to act and their recording of it," and "subjective awareness of decisions is very unreliable . . . it's difficult to pinpoint the moment at which you make the decision." Perhaps the most serious issue is related to the assumption that the time of the measured electrical activity increase—called "readiness-potential" by Libet—is when the brain conceives the "unconscious" plan to push the button. This ignores the very real possibility that the "readiness potential could just relate to the act of paying attention to the wrist or a button, rather [than] the decision to move." In fact, a later experiment tested this alternative explanation with a slightly modified design using two buttons and images on a screen and found that "participants showed 'readiness potential' even before the images came up on the screen, suggesting that it was not related to deciding which button to press."[22]

I would conclude that this bit of science behind Harris's case against free will is just another example of the misuse of science to try to prove a false ideological claim.

Agency, the action of a being, depends on will. You need to decide to go shopping before you act to pick up your keys and wallet. This is why some atheists are so adamant about the nonexistence of free will—if we actually have free will, then there has to be something else in the universe besides deterministic, materialist causes for everything, including for what we humans do. Consistent deniers of free will would say, for example, that I didn't decide to write this book for some "purpose" that I imagine—I simply had no choice in the matter because the neurochemistry of my brain and all my previous experiences compelled me to do so.

While this approach to human (and other animal) behavior might be interesting as a parlor game, and perhaps even deserves to be treated philosophically, it has nothing whatsoever to do with science. The absence of free will is much like the existence of a multiverse—it's a proposition that can be argued about but cannot be tested scientifically.

As Justin Brierley wrote in his book *The Surprising Rebirth of Belief in God*,[23] and as I and eleven other former atheists wrote in *Coming to Faith Through Dawkins*,[24] these sorts of extreme ideas put forward by atheist activists, denying the truth of what most people feel is an essential part of their existence, just don't cut it. And many people are beginning to realize that yes, if God is not real, the reductionists have a point—I am simply a flesh-bearing robot, with no actual willpower, purpose, or meaning, and Dawkins is right—at bottom there is no design or purpose, simply the arbitrary laws of physics and a universe of pitiless indifference.

But the fact—indeed, the scientific fact—is that the New Atheists were wrong. Purpose, agency, consciousness, and will are real phenomena, and therefore (by the reverse of their own logic), God must be real.

### *The Source of Agency and Teleology*

How can we scientifically explain the fact that living cells act on their own? How did the most primitive forms of life start acting with a purpose, no matter how simple that purpose might be? What is it about the jumble of chemicals in a membrane that gives it the power to do something that nothing else in the universe can do? In other words, where does agency come from? Some of the answers are in fact related to what must precede action—the will to act, which in turn is a result of cognition.

The relationships between agency (the will to act) and instincts, neurophysiology, and psychology are complex, as is the entire subject of brains, minds, souls, consciousness, and human cognition in general. To begin to understand these relationships, we must turn to one aspect of life that has made an unexpected and surprising leap into the recent scientific literature with enormous consequences for how we understand the basic nature of life—the reality of cognition in all of life, including the simplest forms.

## Cognition

Cognition can be defined in many ways. One useful description of cognition is "the mechanisms by which animals acquire, process, store, and act on information from the environment. These include perception, learning, memory, and decision-making."[25] This definition focuses on the mechanisms by which cognition occurs in animals.

Pamela Lyon, a microbiologist, extends this to cover any living creature:

> Biological cognition is the complex of sensory and other information-processing mechanisms an organism has for becoming familiar with, valuing, and [interacting with] its environment in order to meet existential goals, the most basic of which are survival, [growth or thriving], and reproduction.[26]

We tend to think of acquiring, storing, and processing information as being components of human or animal intelligence, but as we will see, they apply to far simpler creatures as well.

Some of the skills that accompany being cognitive include the ability not only to move but to move *with purpose*—in other words, to navigate. Cognition allows a creature to learn from experience and to communicate and coordinate activity with other individuals of the same species.

It isn't just humans who are cognitive creatures, and our fellow cognitives aren't just whales, elephants, wolves, or bears. There is evidence for cognition throughout the animal kingdom. This might not surprise pet owners, but the following tends to surprise most people: it isn't only the animal kingdom that shows all or most of the signs of cognition. It's actually *all* of life, including single cells like bacteria, slime molds, and fungi and plants of all kinds.

How is this possible, since so many of the characteristics of cognition are associated with nervous systems and brains, which are found only in animals? The answer is that while brains are great at doing most of the things that cognition requires—sensing, remembering, learning, and so on—having a brain is not the only way to be cognitive.

*Cellular Self-Awareness*

How is it possible for all of life, including brainless bacteria, to be cognitive? One clue is provided by William B. Miller Jr., one of the pioneers of a new field called cognition-based evolution (CBE).[27] Miller says that in order for a cell to act, or to have a reason to act, it must first be self-aware. In other words, it must be aware somehow of its situation, its status, its condition—in a word, of *itself.* I know this is starting to sound surreal, but that's only because it's new. Self-awareness of single cells like bacteria is not crazy, and it doesn't mean that cells spend time in a psychologist's office discussing their motives and experiences. Cellular self-awareness is a fact for all living cells, whether living single like bacteria or as part of a plant or animal. Yes, those dying skin cells that come off your body were aware of themselves while still alive, as is every other cell in every organ and tissue that make you you.

That is what cognition is all about—acquiring knowledge and understanding through awareness, which is the sum product of thought, experience, and the senses.

*Cognitive Abilities*

An organism must have certain abilities in order to possess some degree of cognition. These include the ability to sense and recognize important features of the environment; to assign some sort of value to the information it gains; to adapt to this environmental information as appropriate for its survival and well-being; to retain information about the past (some form of memory) and adapt behavior according to those memories; and to be able to predict future events, at least in the short term and based on information from past experiences.

Many single-cell organisms have all these capabilities. In addition, some have the ability to internalize and combine information from different sources and to interact with other cells of the same kind in order to take part in collective action. This involves the ability to tell the difference between "us" and "them."

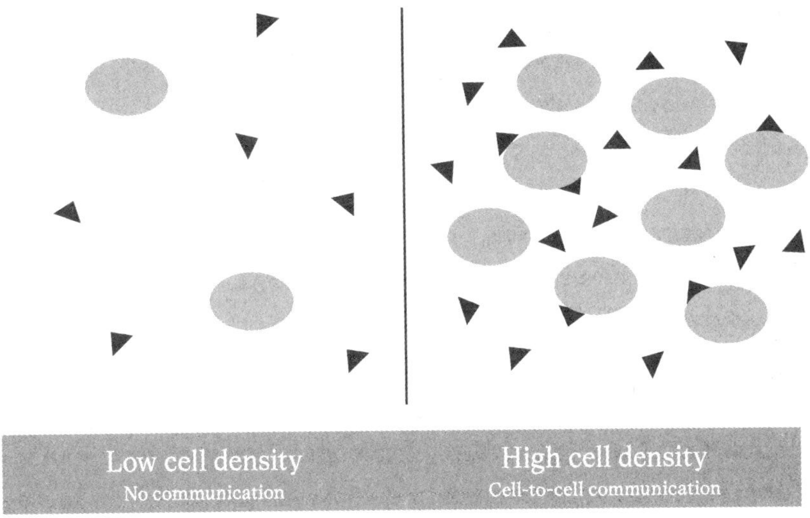


**FIGURE 4.1** *A single cell does not have the capacity to produce light on its own. But many cells together in a clump can produce a useful amount of light. How do the cells know there are enough of their friends around them so that it makes sense to activate the light-producing mechanism? Those little triangles are chemical signals from each cell that attach to receptors on the other cells' membranes. When enough of them do so, each cell "knows" that it is surrounded by friends. They then turn on their own individual lights, which together shine through the darkness.*

### *Signal Transduction as the Mechanism for Cell Cognition*

All cells access information about themselves and their environment. But they don't use any of the kinds of senses we use (sight, sound, smell, touch, and taste) to gather this information. Cellular cognition doesn't work with the neurological mechanism based on electrical impulses, nerves, and brains that we find in animals. Instead, it is based on a well-known biochemical mechanism that has been studied in fields such as endocrinology, organ physiology, and many other aspects of biochemistry called signal transduction (ST).[28]

The actions of hormones on many organs of your body are good examples of ST. A hormone is a complex molecule, often a protein, that has a particular shape unique enough to specifically bind to an equally complex molecule (usually also a protein) called a receptor. When adrenaline (a hormone) is produced in the brain, it travels through the bloodstream and finds receptors it can bind to on various tissues and organs. These include the heart, lungs, liver, muscles, and skin. Adrenaline, like all hormones, doesn't actually do anything other than bind to these specific protein receptors on the outside of cell membranes.

What happens next is the show. That receptor protein transforms when it binds to a wandering molecule of adrenaline. It changes its shape, leaves its watchful post on the membrane, and goes into the cell, where it can do all kinds of things depending on the cell type. It might bind to several other proteins, which together enter the nucleus and interact with a specific set of genes to turn them on and start a chain reaction that will affect the physiology of the cell.

The effect of this biochemical activity in cells is a change in the physiological state of the organ. For example, the heart will start beating more rapidly. The lungs will react by drawing in more air, and the skin will produce more sweat, which will evaporate, thus cooling the body, whose temperature has risen from increased activity. The muscles will be able to perform longer and with more strength than usual, and the liver will release more sugar to power all this activity. All of this takes place without any involvement of your conscious brain.

When you are walking in the woods and see a bear, you might feel fear, but you are not consciously telling your heart to beat faster. It's the part of your brain that is connected to your senses (vision, in this case) that tells millions of cells in many organs of your body to do what they must for you to stay alive in this situation. And the brain tells this to so many different organs by releasing a single molecule—the hormone adrenaline—into the bloodstream.

*Examples of Signal Transduction in Cognition*

That is how signal transduction works in animals, but it also works in single cells with none of the sensory capabilities we have. Signal transduction came first, and the senses of animals came later to improve and build upon the more basic mechanisms of awareness. And the result is not just self-awareness but also awareness of the outside environment, including the presence of others (as shown in figure 4.1).

The results of this ST-based cognition on the cellular level are astonishing. The following quotes from peer-reviewed scientific literature are based on experimental research into how cells function using their cognitive abilities:

> Cells are cognitive entities possessing great computational power.[29]

> Cells are essentially autonomous, sentient and ingenious. . . . They create and regulate activity, respond to current conditions and . . . take decisions to deal with unforeseen difficulties.[30]

> During predation of other bacteria, thousands of *M. xanthus* cells coordinate their behavior to lyse and degrade prey. . . . Cells solve the challenge of migration by migrating together.[31]

> Slime molds . . . include behaviors like problem-solving skills and the ability to learn. . . . They also display the ability to anticipate environmental changes based on prior experience.[32]

The papers those quotes are from provide plenty of experimental, empirical evidence for the cognitive abilities of bacteria and other primitive organisms. Here is another example.

The common bacteria *E. coli* has more than ten thousand receptors on its plasma membrane, many with multiple binding sites for the chemical signals coming from the environment and other organisms. These receptors are the bacterial equivalent of the eyes, ears, and noses of higher animals. They provide information to the cell about its position and status and the potential for danger or for finding food.

Even more striking is that these bacteria and others have clusters of these sensing receptors at one end of the cell. It appears that the components of these clusters can interact with each other, in analogy to the way neural cells interact in brains to process complex information arriving at the cell in the form of chemical signals. These clusters have been linked to purposeful, directed cellular motion. The term *nanobrain* has been used to describe the receptor clusters.[33]

The ST systems of bacteria allow them to sense and respond to an enormous number of signals of environmental conditions. These include important chemicals such as amino acids, sugars, oxygen, and phosphate, as well as physicochemical conditions such as acidity, temperature, and light. They also receive biochemical signals from other living or dead cells (like ammonia and DNA fragments).

### *Biochemical Cognitive Mechanisms*

Research has revealed that the detailed chemistry of how the signal transduction systems work to provide information to the cell is quite complex. This is not surprising, given that these systems support complex cognitive processes.

Some of the receptors that bind the environmental signals are enzymes. When the binding between one of these receptors and a particular signal (like an amino acid) occurs, the enzyme catalyzes a chemical reaction that adds a phosphate group to itself. This added phosphate is then transferred to another protein in the cell, which acts as a marker for a particular state of the outside world (e.g., "there are amino acids near me"), and it can start a cascade of enzymatic activity that eventually results in the chemical activation of a protein that is able to interact with the genome and cause the expression of a specific set of genes. These could include those involved in motion and changes in the structural components of the cell to allow for incorporation of food (like amino acids) across the membrane.

### *Cellular Learning*

The idea that bacteria can learn and act according to acquired knowledge has received direct experimental support. When cells are grown in the presence of a high level of a compound critical for their survival (like phosphate or sugars) and then switched to an environment with very low levels, they respond using their signal transduction sensory machinery and changes in gene expression that allow for survival in the new, difficult environment. But this takes a certain amount of time, and many cells will die.

When the original good conditions are restored, the colony of bacteria will turn off the emergency response systems and recover. If the experimenter then repeats the experiment and switches the bacteria to the nutrient-poor milieu, the same process of sensation, regulation, and adaptation will take place, only this time the process is faster. With every repeat of the experiment, the colony responds faster and with better results. This is due to a kind of learning and memory that can be traced to specific chemical events within the cells that are the legacy of earlier traumas. If we were to wait long enough between repeats of the stress (removal of the required compounds), we would find that the memory is lost, and the colony goes back to the original, slower response.

### *"Smart Cells"*

Bacteria also have biochemical systems that allow them to use current conditions to predict and deal with future events. They can learn, for example, that a change in temperature under certain conditions is followed by a change in oxygen levels. Then, by measuring temperature, they will make adjustments to adapt to new oxygen levels before those changes actually take place.

How bacteria are able to integrate information from many different signals is unknown at present. It is known that cells can merge information coming from different pathways, but the mechanisms explaining how they do this are not yet well understood.

Cognition allows cells to function at an optimum level, both as individuals and as community members. In addition, cognition might direct evolutionary processes. The complex mechanism of protein construction by use of a genetic code and a mutable DNA sequence is able to produce designs that favor survival and reproduction. This suggests that the concept of biological design could derive from the cognitive intelligence of the living cell, but the ultimate source of cognition in primitive living cells remains a mystery.

### *Cognition and Multicellularity*

One of the most difficult problems in evolutionary biology is the origin of multicellularity. Current data suggests that life originated quite early on the new earth, in the neighborhood of a few hundred million years after the planet became hospitable to life in terms of temperature and atmosphere. Then, once the first cells appeared and began to evolve, it took about three billion years before the first multicellular organisms appeared. It's interesting that it took so much longer (about 10 times) to get multicellularity, even with evolutionary processes already present, than to get the beginning of life from chemistry, before biological evolution was even possible.[34]

It is entirely possible that the presence of evolutionary mechanisms after the origin of life might have played a role in the apparent difficulty life had going from single cells to multicell organisms (creatures composed of many cells). Life's unique feature—high-accuracy self-replication—makes stability more common than radical change. And evolution by natural selection means that each living entity has a fitness that determines its survival and chances for propagation. This leads to competition between members of the same species, especially when food and other resources are scarce. Therefore, as time went on, and evolution led to better and better survival mechanisms, cells improved their chances by acquiring characteristics that gave them an advantage over other individual cells. But to become part of a multicelled creature, the component cells that make up the organism had to abandon such competitive strategies and be able to live together in mutual harmony.

In other words, a fundamental component of the evolutionary mechanism, the meaning of "self," had to change from being the single cell itself to being a part of the entire organism as a community of individual cells.

This kind of change appears to be close to impossible to accomplish in a single cell. How could a cell change the underlying selective advantage mechanism for how evolution works? This question has made the rise of multicellularity one of the most difficult problems for evolutionary theory. Perhaps we shouldn't be surprised at the three-billion-year time span it took to get there, considering the enormous challenge required for it to happen at all.

But it *did* happen, and the answer to how it happened might have a lot to do with what we have been discussing in this chapter.

I propose that the origin of multicellularity may be a direct result of the development (probably through a long evolutionary process) of cellular cognition. What does cognition have to do with multicellularity? Everything.

### *A Hypothesis on the Origin of Multicellularity*

As time passed from the origin of cellular life, and evolutionary selection did its work, cells began to develop all the properties of signal-transduction-based cognition discussed above. Part of that cognitive ability was awareness of self and others. Cells gained the ability to recognize other cells that might be enemies or competitors, as well as other cells that are closely related to them and potentially able to join with them to accomplish a common goal, such as the production of light and toxins. At some point, single cells were able to physically join together in a cooperative structure in order to achieve a purpose for a limited time period.

A good example of this, which can still be found today, is slime molds. There are many kinds of slime molds, scattered all through the various branches of life. The most interesting ones are from the group called Dictyosteliida.[35] They spend most of their lives as discrete, single-celled microorganisms, a kind of amoeba. But if there is a shortage of food, they will join together into a visible sluglike multicelled organism. Then, in a remarkable display of cooperativity, some of the individual cells will turn into spores or seeds for a new generation, while others sacrifice themselves to form a rigid stalk allowing the spore cells to be lifted high enough to catch the wind and disperse.

Now we can begin to see that the development of cellular cognition, and the recognition of similar neighbor cells, can lead to a previously unknown biological phenomenon: cooperation of independent living entities. Though it's common to conceive of evolution in terms of competition, we now know that cooperation is just as important in evolutionary dynamics,[36] and the possibility of cooperation requires the presence of cellular cognition. After all, you cannot cooperate with another entity until you can recognize its existence—and recognition requires cognition.

### *Hypermutation and Cognition*

Another sign of cognition and cooperation in a population of single-cell organisms is the phenomenon of hypermutation as a defense against starvation or toxic attack. Hypermutation is a state in which cells begin to divide in a way that increases the error rate in DNA, leading to a temporary and localized increase in mutations. Since most mutations are lethal, the result is a huge increase in cell death. But since a few mutations have the chance to improve the capability of a cell to overcome the crisis, those few (maybe five or six cells out of millions) will survive and repopulate the colony, avoiding extinction.[37]

The hypermutation process can only occur if the entire population undergoes the rapid increase of mutation frequency *together*, which means there must be communication between the individual cells to signal it's time to activate the error-producing enzymatic pathways that lead to a high mutation rate. This process does not serve the interest of most individual cells, only that of the population as a collective, an example of the kind of self-sacrificial cooperation needed for multicellularity.

I do not know if this hypothesis concerning the role of cellular cognition in the origin of multicellularity after billions of years of single-cell evolution is correct or not. But it is a reasonable hypothesis that can be tested.

You might have noticed something between the lines in the above discussion. It was focused on the concept of cognition, but it also implicitly included the ideas of teleology and agency. After all, to cooperate implies a reason for doing so, a reason that is clear to each member of the collective. It also implies the ability to *act* in a cooperative manner, when appropriate. So we can see that *agency*, *cognition*, and *teleology* were all essential ingredients in the eventual development of multicellular organisms like fungi, plants, fish, birds, and other animals—in life as we know it.

## Conclusion

In *How Life Works: A User's Guide to the New Biology*,[38] Philip Ball makes many of the same points about the universal presence of teleology and agency in life. He demonstrates that mainstream, nonbelieving scientists are beginning to agree that biology is revealing a revolutionary truth about how life works. Ball's explanation for how purpose came to exist within life is that it evolved. As we saw in the previous chapter, evolutionary explanations can work only *after* living cells have acquired the ability to self-replicate, which leaves the question of the origin of biological purpose open to theological investigation.[39]

Hopefully I have made a good case that agency, cognition, and teleology (ACT) are present throughout life and that biological science ignoring these universal features of all organisms is foolish and anti-scientific. But even supposing the case is convincing, what can we do about it? How can we incorporate these concepts into new theories of biology, and what questions are they likely to address? The only answer is that, at this point, the scientific community has no idea. I believe that once the biologist community as a whole accepts the importance of ACT, answers will be rapidly forthcoming. Of course, as is the norm in science, there is a good chance that some of the ideas I have put forth here will be proved wrong. But (again, as usual) eventually correct answers will be found. The result will be the development, testing, and widespread approval of new theories, and perhaps even of some long-desired but elusive mathematical laws.

ACT, of course, also points beyond science. One thing we should not lose sight of in all this is that ACT is different from many of the other breakthroughs in biology. We should remember why ACT was rejected in the first place as being "too metaphysical" or "not scientific." Some might even suggest that ACT uncomfortably points too directly to something beyond the known laws of physics and chemistry and it is outside of the modern scientific consensus. Of course, the same thing could have been said about the realities of quantum theory, cosmology, and other physical science advances (and it was indeed said at the time). In biology, we have some atheist ideologues who even claim that the genetic code is not really a code.

Again, if purpose and agency are baked into life, it's legitimate to ask where they came from, and if cognition is universal, what could have given rise to it? These are questions without answers within established science, but I strongly believe that the incorporation of ACT into the science of biology will lead to even stronger evidential pointers to a creator, an entity we call God, with his own purposes, his own ability to act, and his own cognition at a level we could never begin to fathom.

CHAPTER 5

# Abiogenesis: The Conundrum of the Origin of Life

LIFE IS NOT SOMETHING WE CAN CREATE IN A LAB, nor is there yet an adequate theory to explain the leap from a lifeless world to one that is teeming with life. We know that life must have begun in cells with many of the biochemical systems required for accurate cellular reproduction already in place. But we have no idea how that could have been possible. There is no greater conundrum in biology than the subject of this chapter. Does the origin of life from dead chemistry demand a creator?

## The Scientific Conundrum

Life should not exist. Why do I say that? What we mean by being alive appears to be, in a sense, impossible. When we examine the phenomenon of life strictly from the point of view of physics and chemistry, we find that so many of the basic laws of these physical sciences need to be stretched and exaggerated to the point of absurdity when applied to how life could have begun.

I'm not saying that biology disproves chemistry and physics. No physical laws are actually violated by life, and one can easily account for all the quirks of life according to chemical principles. The known laws of physics and chemistry, however, cannot explain the origins of the complexity of biological machinery like the ribosome, the cell membrane, the genetic code, and even basic enzymatic catalysis, all of which came into being between the time of the origin of life and the time of the last universal common ancestor of all currently existing life forms, before evolution was possible.

Imagine a time traveler from 1875 being confronted with an automobile. He would likely think that this thing that could move at such high speed and without any horses or other visible means of propulsion must be miraculous. But once you explained the workings of the internal combustion engine and how gasoline supplies the power that moves the contraption forward, he would marvel at the ingenuity required to make it but also understand that the laws of thermodynamics and ordinary mechanics still hold for such an incredible machine. But if you told him that automobiles are as natural as stones or trees and can arise spontaneously from a mixture of metals and plastics, he would have as hard a time believing that as we would.

The problem with life is not in understanding how it works using chemistry, but in understanding how it came to exist without a willful engineer or builder. The spontaneous construction of a working airplane from a junkyard of parts is a well-known analogy for this problem. Or we can imagine a whole factory appearing in the absence of human intervention. Airplanes and factories don't violate any laws of physics, but to account for their assembly or construction—in other words, their origins—we need something more: people.

This argument is a favorite of intelligent design proponents and creationists, and for good reason. It generally does *not* apply to how living things of all kinds function, from the human eye to the bacterial flagellum, because evolutionary theory can account for how these organs got to be the way they are. Once living creatures with the ability to accurately reproduce themselves exist, evolution by natural selection follows automatically (see chapter 3).

If we think of life as an automatic result of physical laws (as are hurricanes, volcanoes, and supernovae), there is something missing, like our time traveler confronted with the automobile. And the technology of life is far more complex than any human construct.

So let's have a deeper look at whether my claim above that "life should not exist" is true.

## The Science of Abiogenesis

The field of abiogenesis (the origin of life from nonlife) is very large and active, with hundreds of papers being published every year. It is a field of many subdivisions and many hypotheses but no universally accepted theories—even the question of how much progress has been made, if any, is controversial. Many hypotheses have been proposed but none have made it to the status of a theory, and none have been validated to the point of allowing for real progress. What we do know is that life is made of chemicals, and therefore its origins must lie in chemical processes of some kind. But so far that knowledge has been of very little help in solving the mystery. Application of all our knowledge of chemistry has gotten origin-of-life scientists to only the most basic, elementary steps along the pathways leading to a living cell.

Abiogenesis is one of those areas of biology most strongly crying out for breakthroughs from new approaches. There is a consensus that many years or decades, and probably a lot more innovative thinking, will be needed to reach a satisfying scientific answer to the question of how life could have gotten started.[1]

We cannot make progress without finding the missing concepts of how the machinery required for life could emerge in a world ruled only by the laws of chemistry. Part of the problem is that the question isn't actually just one question—"How could life start?"—but implies a long series of questions. There are more questions than answers, but it is important to be precise about the questions since they provide a blueprint for what we need to discover.

In order to understand abiogenesis, we need to unravel a long process that led from a world of chemicals to the organism we call the last universal common ancestor (LUCA) of all of life. This hypothetical creature was a living cell enclosed by a membrane, complete with DNA, many proteins, a genetic code, a complex metabolic machinery, energy conversion reactions, and so on. LUCA is hypothetical because we have never encountered such an organism, but we know one must have existed from the common features that all life has today.

LUCA wasn't the first living cell. It was the cell, or population of cells, from which all current life descends, and we can assume it was preceded by simpler cells. But even the simplest cell that biologists can imagine—one without DNA or modern systems of information, energy conversion, and metabolism—must have been chemically complex. It needed to have a membrane barrier to distinguish self from outside self and a host of metabolic cycles and energy-producing and synthetic pathways in which all the required chemical reactions are driven by highly specific, efficient, and rapid catalysts. And then we are faced with the problem of how such a complex but still relatively primitive cell could have led to LUCA before evolution based on high-accuracy self-replication was possible (see chapter 3).

At the early part of life's origin, we need to ask questions about how the chemical building blocks of life—the amino acids, the nucleotides, the sugars, and the lipids—were formed in sufficient concentration to be assembled into the large chemicals called polymers that are a requirement for life. And then, how did such an assembly take place? Were the conditions on the very early earth appropriate for the required chemical reactions? In the absence of proteins, what catalysts allowed for synthetic chemical reactions to occur at the speed necessary for life? Where did the energy needed for these reactions come from? These and other difficult questions are discussed further below.

I am not suggesting that any of these issues are unsolvable. In fact, answers to some of these have been found, at least theoretically, and often with experimental support. Even if further research and theoretical work comes up with answers to many of these questions someday, we will still be left with another question: Why is it so difficult to understand the origin of life?

If all this sounds a bit overwhelming, it should. Perhaps the best place to start is where we know life must have come from: the world of chemicals.

## The Chemistry of the Origin of Life

There is no scientific controversy about the fact that life is based on chemistry, and in particular, the chemistry of carbon. The carbon atom is uniquely able to form an almost unlimited number of different molecules, making it truly special when it comes to life. No other atom could be the basis for the enormous chemical diversity needed for biological processes. I will not discuss speculations of potential alien life not based on carbon chemistry other than to say I believe they're wrong.

The facts and laws of chemistry are consistent with the way life works. For example, chemical science explains why a nucleic acid polymer is the best of all possible chemical entities to create an information storage molecule. The fact that each of the units of DNA (the bases or nucleotides) can form weak but specific interactions with only one of the other three units means that the polymer can be copied with very high fidelity. At the same time, there are no restrictions on the sequence of bases on any strand. This means that all possible sequences are possible, allowing for the possibility of the sequence operating as a code. (See the appendix for more information on how this works.)

Furthermore, the chemical diversity of amino acids, along with their ability to polymerize via peptide bonds, make proteins the ideal catalysts. In this way, organic chemistry (the chemistry of carbon compounds) blends neatly into the basic tenets of biochemistry (the chemistry of life). But none of this explains how molecules like DNA and enzymes came into being.

Textbooks on organic chemistry and biochemistry thoroughly describe how the molecular nature of life functions according to the rules and principles of carbon chemistry. But you will find very little in these textbooks concerning the chemistry of abiogenesis, other than perhaps brief summaries of current hypotheses that rapidly become out of date.

This is because so little is known to the extent and standard required for inclusion in textbooks. The rules of chemistry—from thermodynamics to kinetics to mass action—do not help us when it comes to understanding the origin of life. Some of these rules actually work *against* the genesis of the very strange state of chemistry we call being alive.

Left on their own, chemicals do not polymerize. Molecules much larger than a few atoms tend to break up into smaller fragments, a process called hydrolysis when it occurs in water (which is, as you will see throughout this chapter, where life had to begin). Synthetic chemist James Tour has discussed this and other problems affecting progress in origin-of-life research in numerous talks and presentations. I had the honor of appearing with him on one of his YouTube podcasts,[2] and we soon found ourselves discussing the problems with the chemistry of abiogenesis from two different angles. Dr. Tour was focused on the earlier chemical reactions needed to synthesize some of the simple and complex molecules necessary for any kind of life. As a biochemist, I concentrated on how those complex molecules, once formed (somehow), could work together to form the machines and systems required to accomplish the tasks necessary for life. This illustrates that at every stage of abiogenesis, there are unsolved chemical problems.

If we compare chemical reactions in living cells to those that occur in a physical environment like seawater or a warm pond, the contrast is starkly obvious. In biology, chemical reactions are catalyzed by enzymes, so they are rapid and can occur in thermodynamically unfavorable situations requiring energy input. In the absence of catalysts, most chemical reactions are very slow and only follow thermodynamically favored pathways. Catalysis occurs outside of life, of course. There are nonprotein catalysts, like clays and metals, but they are orders of magnitude less efficient than enzymes and usually cannot drive energy-requiring reactions.

## The Energy Issue

I have said that evolution is logically unable to account for complex processes needed for the origin of elementary life. One possible exception is the use of energy. The main source of energy for life today comes from the sun, through photosynthesis, but it's unlikely that the very first life forms would have been photosynthetic. The process is complex, and it would have taken some time to evolve, and that evolution (like all evolution) had to happen in already-living cells that could reproduce and follow the rules of biological natural selection, all of which require energy conversion systems to be already in place.

There are, however, other potential sources of energy on the earth in addition to solar light, both currently and in the distant past. One of the most prevalent is heat. The earth's fiery inner core holds a staggering amount of heat, some of which is released into the ocean in the form of volcanoes and hydrothermal vents. There are colonies of living organisms clustered around such vents in the deep ocean, in complete darkness, which use this heat as an energy source today, so we know this could have happened on the early earth as well.

The efficiency of energy conversions starting with heat is a fraction of that obtained by photosynthesis using light, but it's sufficient for rudimentary single-celled life forms and may have been the initial energy source at the origin of life. If life began near hydrothermal vents in the deep seas (a widely held hypothesis), and heat rather than light was the first energy source, one can imagine a scenario of a slow evolutionary path in cells powered by the simpler energy conversion of heat. In this scenario, living cells could have evolved with heat as the energy source to the point where they could produce the pigments and primitive chemical reactions necessary for full-blown energy conversion from sunlight. The details of such an evolutionary switch from heat to light are unknown and represent yet another complex mystery needing to be unraveled. But it seems that, unlike the systems involved in high-accuracy self-replication, energy conversion is one of those properties of life that could possibly have arisen before the origin of evolution.

## Abiogenesis and Purpose

What about those complex molecules needed by life—is it really that difficult for them to arise? It is certainly true that we can make large polymers in our laboratories if we want to, and we can even make some under fairly primitive conditions if we are careful enough. But the key phrase here is "if we want to." Assuming an absence of teleology in nonliving nature, who or what provides the desire to make polymers? And who or what would have the ability to act on that desire to do so? Nothing, of course, outside of a creator.

Without purpose or agency, there is no reason at all that a pond or lake containing lots of catalytic clay crystals, along with ammonia, carbon dioxide, and maybe even some amino acids and various ringed carbon compounds, would suddenly start producing dipeptides, sugars, fatty acid chains, or anything else. There is no reason for that to happen spontaneously, and if some freakish accident led to a peptide bond between a pair of amino acids, it would be broken rapidly by the hydrolytic action of the water.

But it is even worse than that. The implication from the two previous paragraphs is that if there were some sort of teleology and agency present on the early earth, perhaps due to some yet-unknown, putative chemical law as discussed in the previous chapter, then we might have our answer to how life began. But even that seems dubious—because the experiments have been done, and they are a failure.

We humans are filled with a sense of purpose, we have all the agency and cognition we need, and we can make almost any chemical we want to, including all the chemicals needed for life. But with all that, we still cannot produce a living organism of even the simplest kind. We can either make from scratch or purify every protein, every nucleic acid, every polysaccharide, phospholipid, and everything else that is needed for a cell to live, and we can put them all into a small vesicle. And nothing happens, even under the ideal conditions of a laboratory. This is why I say that the experiments have failed.

I am not worried that by the time this book is published, life will have been created and my comments here will be obsolete. The reason for my pessimism (or perhaps optimism, when it comes to the fate of this book) is that the scientific community has no idea what we are missing, but we do know that something *is* really missing.

## Abiogenesis and Biological Evolution

Most scientists agree that the process by which life evolves once it exists is fundamentally different from the process of abiogenesis. This includes those who strongly believe that both processes are purely natural. In his book *The God Delusion*, Richard Dawkins states, "The evolution of life is a completely different case from the origin of life because, to repeat, the origin of life was (or could have been) a unique event which had to happen only once."[3]

There are, of course, very good reasons for excluding biological evolutionary mechanisms from explaining the origin of life. As I have argued in chapter 3, much of the complex chemistry and biochemistry needed to start life are also needed for evolution to work—especially everything related to accurate self-replication.

For example, invoking biological evolution to explain the existence of the genetic code and the method of making proteins in living cells would be a little like trying to explain the existence of a factory by pointing to a car. Yes, the factory was required to make the car, but no car ever contributed to the construction of a factory. The machinery within the factories of living cells makes everything that the cell needs. All the products of these machines are required for biological evolution to work, so they could not have been originally made through evolutionary processes.

Daniel Dennett has correctly characterized natural selection as a "crane," with special abilities to bring forth the chemically unlikely scenarios of increasing complexity in the chemistry of life.[4] Cranes can construct the buildings on which they are mounted. But if we cannot invoke the crane of natural selection, then we must look for other explanations of how chemistry could have possibly become alive.

## Chemical Evolution

The only alternative naturalistic answer to this mystery I have seen is that some other, non-Darwinian selection-based mechanism was at work to allow for all the components and interactions of complex biological machinery to come about.

Other kinds of evolution based on selection are known. We speak of the evolution of everything from music to cell phones to computers. The myriad technological and ideological changes we can think of throughout history are examples of cultural evolution. For the origin of life in the absence of humans, there can be no cultural evolution, as well as no biological evolution, so such a mechanism would only be relevant if the earth were originally populated by angels with good, working laboratories (an interesting idea, but quite difficult to support scientifically).

There is the possibility of chemical evolution, whereby certain chemicals might become more numerous or even more complex depending on differential stability, relative concentration, and various reaction kinetics of competing molecules.

That idea faces the problem of how to define chemical evolution related to the origin of life. Chemical evolution is totally independent of inheritance-based natural selection. While it is possible that some forms of chemical evolution could lead to the existence of certain molecules needed by life, chemical evolution cannot, even in principle, come close to explaining any of the biochemistry of even a rudimentary living cell.

For example, laboratory synthesis of simple sugars like ribose, a key ingredient of DNA, is extremely difficult and can only be done under tightly controlled conditions that would not be found in nature.[5] So how could such molecules evolve in an abiotic world? What possible processes akin to mutation, natural selection, comparative reproduction, survival of the fittest, and so on could possibly exist in a purely chemical world?

Chemistry has no teleology. There is no final cause of selection to assist chemicals in lasting longer or becoming more complex. There is no natural selection in chemistry that serves any purpose at all. Some chemicals indeed last longer than others; some are more stable or more resistant to hydrolysis than others. But their relative longevity is unrelated to anything that characterizes life. In fact, the most stable molecules tend to be small rather than large, and nonreactive rather than easily reactive with other chemicals. Both of these properties work in the wrong direction for life, and, if anything, one would expect the rules of chemistry to rule out the possibility of life rather than encourage it. Hence my provocative declaration at the beginning of this chapter that life should not exist.

## Biochemical Issues in Origin of Life: Borders and Mass Action

It is useless to try to conceive of life starting outside of a physical boundary, such as that provided by modern membranes. This is due to one of the most fundamental laws of chemistry—the law of mass action. This law states that the extent and rate of all chemical reactions depend on the concentration of the reactants (the chemicals that bump into each other and then change the way their atoms are bonded together). Concentration simply means the number of molecules of a particular type present in a particular amount of space. In this case, molecules dissolved in water.

The concentrations of chemicals in seawater are very low and were low on the early earth as well, so reactions would have been exceedingly rare. Even when they did occur, the chances that the products of those reactions would be available for any further reactions would be remote, given the propensity for molecules to move freely throughout an aqueous environment.

The solution to this problem would be to enclose a bunch of molecules in a space where their movements are restricted so they are not free to diffuse all around the wide ocean (or lake, or pond). We know that such enclosures exist and can form naturally and spontaneously in water. Molecules without charge, or "polarity"—such as fats, lipids, and hydrocarbons—can spontaneously form themselves into capsules or vesicles, trapping other molecules inside them. These structures form a boundary between *inside* and *outside* so the chemical concentration is greater inside than outside the boundary. Soap bubbles are a good example of this natural formation of boundaries in water, with a potential for attracting other nonpolar chemicals, like those in dirt.

The formation of such boundary structures in water is often postulated as the way the first membranes—the first borders for what would become living cells—came about. Such a thing seems at first glance entirely reasonable, and it fits the criteria of a phenomenon that occurs spontaneously and naturally, following the laws of chemistry.

But things are not so simple when we dig a bit deeper. A nonpolar membrane composed of lipids or long chains of hydrocarbons cannot work as an enclosing membrane for a living cell, or even for a protocell. While such a border would work quite well in keeping the molecules needed for life inside, it also keeps everything else outside (except for some other nonpolar chemicals that could get through).

That means the nicely encapsulated protocell would not be able to access most of the chemicals needed to grow, repair itself, or use for energy. Actual membranes of living cells, as well as the postulated membrane for LUCA, are far more complicated than a simple nonpolar lipid membrane. They allow a *regulated* flow of material both into and out of a cell. Biological membranes include a host of proteins embedded in the lipid structure that can serve as channels for the passage of salts, metals, and organic compounds of all kinds in only one or both directions. They also have a variety of proteins acting as receptors for signaling molecules from other cells that allow for cell-to-cell communication and cognition.

The functional biochemistry of membranes is a fascinating and complex field that is continually revealing amazing new features of even the most primitive examples of biological barriers surrounding the contents of living cells. Without the functional complexity of a membrane that allows for regulated transport of molecules, life would have no chance of getting started.

## The Metabolic Web

To make a typical biochemical—a molecule needed for life that is made within cells—many chemical reactions need to occur, often in a stepwise pathway, where the product of one enzyme-catalyzed reaction becomes the reactant for the next step. These pathways are complex, and they need to be regulated so that the amount of the final product is not too much in excess or deficit.

Some of these regulatory mechanisms are relatively simple, such as using the final product as an inhibitor of the first enzyme in the pathway to avoid excess. But that degree of regulation is generally only the beginning. For many pathways, there are intersecting and connected signals for activation and inhibition, so that the final balance between synthesis and degradation and no action at all is extremely finely tuned. Add to this the regulatory control exerted by the activation or repression of gene expression that produces the relevant enzymes, as well as the varying energy requirements for the many individual reactions, and you will realize that understanding this biochemical complexity is something like trying to account for all the activities and interactions of all the individuals in a large city.

## Abiogenesis Hypotheses

So far, most of what I have written in this chapter is pessimistic about the possibility of coming to a scientific understanding of the origin of life. But science has always operated in an atmosphere of optimism that answers exist even for the hardest problems, and I believe that despite all these problems and hurdles, there must be a solution to how life began. Here we are, after all, along with lots of other creatures. And while progress has been slow and difficult, hypotheses abound. There are both grand, general hypotheses and limited, specific hypotheses about narrow issues. Some hypotheses have even gained major support from experimental evidence and reached near consensus, only to be struck down by new data. We will address some of these, but first we need to clarify a few fundamental ideas.

Researchers in the field are trying to figure out how life *could* have gotten started, which is more feasible than asking "How *did* life begin?" The approach to answering the "could" question is to propose mechanisms and then test them against the chemistry and biology we know are true. For example, one hypothesis would be that life got started by a lucky accident: DNA and several enzymes spontaneously formed in a membrane-enclosed vesicle, and life began to function. But this is chemically impossible and is therefore ruled out by all biologists.

Here are some of the questions that need to be answered:

- Were the conditions on earth appropriate for the required chemical reactions?
- What were the first catalysts, in the absence of proteins, to allow synthetic reactions to take place at the velocity necessary for life or protolife?
- How do we deal with the known chemical instability of some of the components of biochemicals like cytosine and adenine?
- What energy sources could have been used to fuel the chemical reactions?
- Where did a genetic code come from? How did the translation machinery develop?

As I've said before, all of the above had to happen in the absence of biological evolution.

Origin-of-life research has traditionally been divided into two camps. One group, the "replicators-first" theorists, think that molecules capable of self-replication (like DNA or RNA) appeared first, and then metabolism followed. The other group points out that replication is not possible without enzymes and metabolic processes already in place. Both sides are probably right. Proteins need genes (DNA) in order to be made, and DNA needs protein enzymes to replicate itself. Without enzymes and metabolism, it's hard to imagine efficient and accurate replication. But without replication, any advanced metabolism that arises in a particular protocell cannot survive into the next generation. So far, no theory of abiogenesis is more complete, convincing, or devoid of serious flaws than any others.

On the other hand, there are some promising ideas on the origin of life that are still in the development phase but could provide at least some answers. In recent years, the MIT physicist Jeremy England has developed a hypothesis that life might have been highly favored to begin as the most efficient way to dissipate heat energy from the sun into the surrounding environment.[6] It is an intriguing idea from the point of view of physics and suggests that some kind of primitive metabolic life-form might be a natural occurrence, given the right conditions. It also addresses the issue of hidden teleology behind abiogenesis. But it depends on unknown chemical mechanisms that allow for self-sustaining metabolic autocycles and ignores all the rest of biology, such as the need for catalysis, specificity, activation, and rapid and highly efficient reaction mechanisms. As far as I can understand it, it does little to explain the origin of biological mechanisms that are critical to allow for life to evolve.

## RNA and the Origin of Life

For quite some time, the most widely held scientific theory of abiogenesis has proposed that an "RNA world" of life arose first and then morphed into the modern DNA world. The discovery of catalytic activity by some RNA molecules (called ribozymes)[7] led to a general consensus that the original information-storage and replicating molecules were composed of RNA, in the form of these catalytic ribozymes.

RNA is very similar to DNA in structure and composition. It comes in many forms and can carry out many functions, including sufficient catalytic activity to allow for self-replication. Thus, RNA ribozymes are theoretically able to take the place of both protein enzymes and information-carrying DNA. This led to the hypothesis that an RNA world could be a solution to the dilemma of which came first—replicating information storage or catalytic metabolism. In an RNA world, both activities could be present in a single kind of molecule.

A lot of evidence was gathered to support this theory. The RNA world could include a limited form of natural selection, since the most efficient self-replicating RNA molecules would have a selective advantage and be expected to increase their numbers. If the RNA sequences are accurately copied, and the same sequence has a high degree of catalytic efficiency, then both genotype (the sequence) and phenotype (the efficacy of the catalytic ribozymes) would be inherited.

Following this scenario, it is possible to conceive of protocells containing a population of RNA ribozymes that evolve in size and complexity to the point where they are able to catalyze their own replication. They may also be able to catalyze other reactions, such as the formation of peptide bonds,[8] thus making small chains of amino acids and energy-producing cellular reactions. The RNA world concept quickly became the most viable theory for the origin of life among biologists.

### *Problems with the RNA World*

But the excitement about the RNA world as a major step in the origin of life has faced several stubborn problems. Further research showed that the efficiency and accuracy of RNA replication in the absence of protein enzymes was not good enough to allow for a stable informational state.

The following quotes from the scientific literature illustrate some of the problems that arose with the RNA world once the details were thoroughly examined:

> If emergence of the first RNA replicase ribozyme coincided with the origin of life, it would have had to arise in a single step from prebiotically synthesized RNA, without the benefit of Darwinian evolution. . . . Current understanding of prebiotic chemistry argues against the emergence of meaningful amounts of RNA molecules, even a tenth this length [the length necessary for replicase activity].[9]

> The formation of an information-bearing homopolymer [RNA] through undirected chemical synthesis appears very improbable.[10]

> The preservation of encoded information processing during the historically necessary transition from any ribozymally operated code to the ancestral aaRS enzymes of molecular biology appears to be impossible, rendering the notion of an RNA Coding World scientifically superfluous.[11]

> The evidence that is available at the present time does not support the idea that RNA, or an alternative replicator that uses the current set of RNA bases, was present at the start of life.[12]

Other issues include the tendency of long RNA strands to "self-anneal"—to fold up and stick to themselves or to other RNA strands.[13] Because annealing is about a thousand times faster than spontaneous noncatalyzed replication, left on its own, RNA would probably never replicate itself.

The number and sophistication of proteins or large peptides would be severely limited in an RNA world because each catalytic step in the synthesis of a polypeptide would require its own ribozyme. Such a living system could not evolve to the level of a complex modern bacterium, and certainly not be able to support the degree of complexity in the protein population necessary for even the simplest multicelled organism.

### *Transition from the RNA World to the DNA World*

If RNA-based life-forms had arisen spontaneously and achieved some degree of success, at some point there had to have been a shift to DNA-based life. The advantages of the modern DNA-based life system include the possibility of major gains in complexity following a far more efficient process of natural selection. This is because such a system allows for the production of an enormous array of proteins, including catalytic enzymes, which in turn can assume an astonishing variety of structural and functional roles. In DNA-based life-forms, all the sequence information that was present in multiple, diverse RNA ribozymes is stored in a single informational molecule, the DNA.

The problem is that we do not know of a feasible way in which the RNA world could have transitioned to the DNA world. It is exceedingly difficult to explain the origin of the genetic code through stepwise evolution by natural selection using preexisting components from the RNA world. There is an astonishing degree of complexity in the molecular mechanisms that convert information from the genetic molecule into specific protein sequences, giving the cell all its phenotypic characteristics (see part 1 of the appendix).

Origin-of-life researcher Charles Carter has illuminated problems with the possible transition from an RNA world to a genetic-code-based DNA and protein world. While the arguments used by Carter and his collaborator Peter Wills are exceedingly technical, they are convincing to specialists in the area of translation (the part of biochemistry that deals with protein synthesis and the genetic code). For example, Carter and Wills state in one of their papers, "Spontaneous folding of RNA aptamers and the dynamics of an RNA world do not require encoding into genetic information and hence fall well short of explaining the separation of phenotype from genotype necessary for true Darwinian evolution."[14]

The extreme complexity of the translation system is behind a significant part of the difficulty in coming to any useful conclusions on how modern life, with its high-accuracy self-replication systems, could have come into being. This quote from a paper by Eugene Koonin and Yuri Wolf says it all:

> The translation system might appear to be the epitome of irreducible complexity because, although some elaborations of this machinery could be readily explainable by incremental evolution, the emergence of the basic principle of translation is not. Indeed, we are unaware of translation being possible without the involvement of ribosomes, the complete sets of tRNA and aminoacyl-tRNA synthetases (aaRS), and (at least, for translation to occur at a reasonable rate and accuracy) several translation factors. In other words, staggering complexity is inherent even in the minimally functional translation system.[15]

(I'll mention that neither Charles Carter, with whom I have spoken, nor Eugene Koonin are theists.)

These and other problems have led to an increase in skepticism among scientists about a purely RNA world. It has been recently replaced by a new hypothesis of an "RNA/peptide" world in which small proteins made up of five to twenty amino acids (peptides) were present to assist RNA in its functions, which would allow for possible evolution into the modern system.[16] It remains to be seen if this theory will do better than the pure RNA world in solving the enigma of abiogenesis. At the time of this writing, I think the verdict is best rendered as "probably not."

## Some General Theoretical Problems with Abiogenesis

The default scientific assumption is that the origin of life must have been a stepwise and continuous process and that the earliest living cells must have been quite different from anything alive today. But the idea of a continuous pathway from nonliving chemistry to a living cell has been questioned by a number of scientists, including myself, on both experimental and theoretical grounds.

I have published two papers on the problem of getting biologically evolving life started via a continuous process of slow, stepwise improvements by some kind of natural selection.[17] These papers use a theoretical model based on probability statistics to show that when the accuracy of the replication process is below a certain threshold, evolutionary progress is impossible, and the protocells will simply go extinct. What this means is that life must have started out with many of the systems required for accurate reproduction already in place, but before there was any mechanism to allow this to happen.

Another major theoretical roadblock to finding a theory for abiogenesis is suggested by some recent work by chemist Lee Cronin, astrobiologist Sara Walker, and their colleagues on "assembly theory."[18] The authors define a measure of chemical complexity called the molecular assembly index, and their results indicate that there's a threshold of complexity above which only chemicals that exist in biological systems are found. They conclude that in the abiotic world, compounds like ATP (adenosine triphosphate—the universal biological carrier of energy) or a simple tripeptide (three amino acids linked together) would not be produced, at least not in significant quantities.

In that paper, the authors discuss this result in the context of biosignatures as a possible tool for finding life on other planets, with no discussion of the implications for the origin of life on earth. However, in a later paper, they state the following:

> It is important to stress that we are not arguing that the threshold cannot be crossed without biology, since a biological system developing from purely abiotic ones has clearly happened at least once. However, this would require systems outside of our model, such as replication, duplication, and evolution. The exploration of the objects and processes that cross this threshold will be of key importance to our understanding of life.[19]

The problem with that statement is that it is self-contradictory. They say that they are not claiming that biological systems are required to cross the threshold from nonlife to life, but then they admit that in order to cross that threshold, replication and evolution are needed—all of which are properties solely of life. My own work and that of Cronin and his colleagues are just two of many examples of scientific findings that point to major hurdles in our understanding of abiogenesis, all of which cry out for a paradigm shift in our thinking about the issue.

## Divine Design

As the bioinformatics theorist Hubert Yockey noted in a 1977 publication, "One must conclude that, contrary to the established and current wisdom, a scenario describing the genesis of life on earth by chance and natural causes which can be accepted on the basis of fact and not faith has not yet been written."[20]

I believe that quote still applies almost fifty years later. We still do not have a working theory of the natural causes that led to the genesis of life.

What are the theological implications of the abiogenesis conundrum? I have stressed throughout this book that "something is missing" because that summarizes well my view of where we stand as biologists in so many areas. And nowhere is this truer than in life's origins. What is missing? New scientific approaches, certainly, but perhaps also an acknowledgment that, like the beginning of the universe, the origin of life seems to demand a creator. Not just an *intelligent* designer (humans are intelligent, and we cannot design life), but a truly *divine* designer. And perhaps the invention of information—which allowed for a genetic code and the subsequent evolution of all living creatures, including ourselves—was due to an intervention by that designer, who is not amenable to scientific investigation.

I do believe that God created life. We could just leave it at that and move on, but I also believe that would be a wasted opportunity. Here is where I feel a strong pull from the spirits of Robert Boyle, Michael Faraday, Louis Pasteur, and Francis Bacon, along with all the other scientists inspired by their Christian faith in the past and present, who believed and acted on the premise that all of creation is a book written to tell us about the glory of God. We already know that God created everything, including life. The challenge we face is understanding how he did it. We have been successful in answering that "how" question for a lot of how the natural world works, and I am hopeful that with God's guidance and mercy we will be able to do it for this fundamental question of biology as well. Our task is to learn how to read and understand that book of works (the world of nature) to learn all we can, not only about the reality of our world but also about its creator.

This means that, while trying to understand how God created life, we are likely to learn many hidden secrets not only about the deep truths of biology but also about God and his purposes. So I would say let's *not* leave it at that. Let's continue to work hard to find the missing laws that will not only explain the transformation of chemistry into living cells but also illuminate some less understood areas of the biology of plants, animals, and of course ourselves. I don't know anything about what those laws and theories will be, or even what they will be like. But I believe that if and when we find them, they will prove to be as breathtaking and beautiful as all the best science, and they will point even more strongly to the glory of God.

In my book *The Works of His Hands*, I stated that not all knowledge can be obtained from scientific work, using the tools and theories of the science of today or tomorrow. Whether that will be true of abiogenesis is still unknown. But I am convinced that it is true for a deep understanding of the nature of humanity—specifically, what we mean by the human soul and by being created in the image of God. The next chapter is all about that.

CHAPTER 6

# In the Image of God: Human Behavior and Consciousness

HUMAN BEINGS ARE UNIQUE AMONG CREATURES. Scientific materialists have sought to explain these differences according to natural selection, but such explanations seem more like "just so" stories without much real science behind them. The complexity of humans' moral behavior and consciousness cannot be understood in an exclusively biological framework, and human art, beauty, and love seem to point beyond a purely materialist approach to the reality of being made in the image of God.

## About Humans

Perhaps the most important and least understood topics in the science of life involve things related to us, human beings. Human nature, including consciousness, the soul, and morality, are just barely subjects for scientific inquiry and might be considered at this time to be in the realm of metaphysics.

Can the methods of science even broach these topics, and if so, how? We are indisputably alive. We are part of the biological realm, and our very existence calls out for a higher view of the nature of life, a view that cannot exclude elements of theology and grace. How marvelous it is that the Lord God created an entity, a part of the natural universe, that can type words on a computer that can be read and understood by other similar creatures. How wonderful that those shared words can evoke in us emotions such as joy, anger, or agreement—and even, perhaps, worship of our creator. We are creatures made of quarks, protons, electrons, and bosons who are writing and reading and feeling and thinking about it all. How amazing this is, and how far outside of any explanation limited to the known laws of physics and chemistry!

We need to illuminate and address the age-old questions: What does it mean to be human? What is the moral sense, and where does it come from? What is genius, creativity, despair, hope, worship, laughter, music, and art? And what is human consciousness?

My book *The Works of His Hands: A Scientist's Journey from Atheism to Faith* includes a chapter called "People" in which I lay out my views about human exceptionalism. I explain that coming to faith in the divinity of Jesus Christ had no major effect on my earlier views on the nature of humanity. Even as an atheist, I had always known that people were special, much more than just another animal of the primate variety.

But once I became a Christian, I was able to connect my feeling about the unique status of human beings to those words in Genesis: "in the image of God he created them," and the reason behind the striking and undeniable elevation of people beyond the world of the rest of biology became crystal clear. My "People" chapter is mostly a defense of this view of humans, which historically had been an integral part of the old style of "humanism" but is now often rejected by the newest brands of atheism and found mostly in Christian and other religious worldviews.

One excerpt from that chapter will make this point, which I will not continue to labor here:

> Yes, we are primates, but somehow we must account for the masterpieces of art, astounding works of literature, mathematics, self-sacrifice, leadership, poetry, internet chat forums, the composition of a sublime symphony by a deaf composer, the articulation of the physics of singularities by a paralyzed scientist, the daily endurance and love of "ordinary" people.
>
> Why do we paint? Why do we make up stuff and write it down? What good is music? And why do I sit down and type on a keyboard to communicate things to be read by people I have never met? I think it's the same reason we ask, "How does that work?" "Why does this happen?" or "What is behind that thing I see?" Science, music, art, philosophy, love, and humor are not necessary for existence—we could live without them. But they are all part of what it means to be human. So why are we the way we are? . . .
>
> The process of evolution by natural selection produced our bodies. . . . But I maintain that our physical nature is not all that defines us, and evolution cannot account for everything that makes us human.[1]

The physiological and biological nature of human beings is well understood scientifically and fits into the evolutionary framework of the Primate order. The genus *Homo* contains one living species, *Homo sapiens*—us—with at least five other species having gone extinct in fairly recent biological history. We ourselves are a baby species, born roughly two hundred fifty to three hundred thousand years ago. There are a few interesting genetic and physiological features of humans—our loss of most body hair and our opposable thumbs come to mind—but what makes us really interesting and unique is our behavior. By that, I mean everything we think and do.

## Human Moral Behavior

One of the many debates about moral behavior centers on the idea that there is a strong genetic component to it all, which would also imply that evolution can lead to different behavior patterns depending on how they affect survival and reproduction.

I believe that the connection between genetics and moral behavior is less than straightforward, to say the least, and frequently overstated. There is, of course, no question that genes have played a role in shaping behavior throughout evolutionary history. The beneficial effects of a behavior like individual sacrifice for the sake of a community, for example, cannot be denied. But is there a high moral value attached to a bee's self-sacrifice for the good of the hive the way there would be in a human group? Does a bee decide, in a moment of supreme valor, to give her life for her fellow bees? I don't think so. I think there are no moral imperatives involved in the bee's case, only instincts that are a result of evolutionary pressure.

Most people would agree that this is obviously true when it comes to the evolution of viruses, bacteria, and various animal and plant predators and parasites. No viruses (such as COVID-19) decided to be "good" at one point and switch to targeting the upper respiratory tract rather than the lungs. Increased transmissibility and decreased host mortality is a well-known and frequent evolutionary strategy for infectious, lethal viruses, and the resulting good news for humans is simply an accident.

While these might be "extreme" examples (bees have a complex social system but no brains, and viruses are barely alive), the same sort of thing applies—to a large extent, though not entirely—to behavioral traits coded for by human (as well as other animal) genes. For example, increased empathy might be expected to result in behaviors that are more "morally good" according to most human cultures. But a careful examination of the literature shows this is not always the case. Various autonomic reactions, as well as changes in empathy, may lead to various behavioral outcomes. It is known that both emotional states and behaviors have multigenic causes and involve highly complex gene-environment interactions. This makes prediction of anything to do with moral actions based on genotypes of one or a few genes extremely difficult.

As I wrote in my chapter of the book *Coming to Faith Through Dawkins*, "Enough studies have been done on a variety of behavioral traits to show that the genetic component of such phenotypes is roughly between 30 and 50 percent, with cultural and other non-genetic sources making up most of the contribution to how people behave. Moral behavior is clearly among those."[2]

A particular genetic variant in the oxytocin receptor gene (OXTR) is an example of this complexity. One paper describes a behavioral phenotype—problem behavior in children whose mothers had post-partum depression—in which the gene variant called the A allele produces a variant phenotype.[3] Another paper connects the G allele of the same gene to a different behavioral phenotype (increased empathy).[4] The two alleles at this locus have also been associated with a variety of other individually tested behaviors.[5] Clearly both alleles have a variety of effects, and neither is necessarily better than the other. It appears from these studies that this polymorphism is in Hardy-Weinberg equilibrium, meaning that selection pressure for either allele is currently absent, and the gene probably was never a target for natural selection.

In addition, the very fact that human cultures differ significantly in their moral systems could be seen as evidence against a strong evolutionary genetic connection. There are only small differences in allele frequencies of some of these genes in different ethnic groups, and these differences are generally not meaningful. Attempts to find such small differences and attribute meaningful behavioral outcomes to them often reflect unsavory motivations. At one point in my career, I was part of a review team that examined several grant applications on behavioral genetics. One application hypothesized that different frequencies of a particular polymorphism in a neurotransmitter gene could make African Americans, as a population, more violent than European Americans. Preliminary evidence was nonexistent, and the proposal was considered by the whole committee to verge on outright racism.

This does not mean that there are no genetic correlates with human acts of moral good or evil. All humans (except for rare variants like psychopaths) recognize the inherent good in caring for children, showing love to one's spouse, and being kind to kin and relatives, and all recognize that doing harm to these people is bad. There is no question that these moral understandings and the resulting behaviors are of evolutionary origin since the same behaviors are seen in many mammalian and other animal species.

But we only need to think of moral judgments in human cultures that extol the killing of "enemies," the mistreatment of people with nonconforming phenotypes, the subjugation of women, and so on to see that once we go deeper than fundamental common values, we find many cultural determinants of morality, with no genetic influence at all.

Cultural evolution, which explains the different moral codes in human societies, is fundamentally different from biological evolution, which depends on genetic variation. For this kind of moral variation, it is memes, not genes, that count. Cultural evolution, which plays an incomparably greater role for humans than for any animal species, has given us everything from our use of fire to the internet and everything else in between that makes us more than just another ape. This includes the moral choices we make.

## Evolution and Human Behavior

Modern atheists often propose to accept as axiomatic the principle that *all* human behaviors, like all the details of human anatomy, are the result of evolutionary adaptations to survival conditions.

Sociobiology and evolutionary psychology are two major fields of scholarship based on explaining human behavior in terms of evolutionary adaptations. There are, of course, plenty of behaviors where an evolutionary link can be found. Things like sex differences in mate choices and kin altruism can be explained as adaptations, with cultural evolution then elaborating and adding variations in cultural mores.

The problem with this method, though, is that it has the potential of veering away from science and into the realm of the "just so" story. One example of such an evolutionary psychology "just so" story published in a peer-reviewed mainstream scientific journal is about why girls prefer pink or reddish hues compared to boys who prefer blue. The conclusion—based only on what sounds logical—is that this color preference is probably due to the practice of ancient females gathering berries, while the males hunted.[6] This sounds logical—except for the fact that female preference for pink is a modern Western cultural development, and less than one hundred years ago pink was considered a masculine color. Clearly color preference is a social construct and cannot be related to any evolutionary mechanism.

I give more examples of these stories in *The Works of His Hands* and conclude that "there is no possible human trait, real or imagined, that cannot be postulated to result from some evolutionary selective advantage."[7]

In many such explanations, there is no actual evidence for the claims made, other than that they *sound* logical. This is not science but a modern way of trying to fill in a real gap in our knowledge of human biology and psychology—not with God, of course, but with pseudoscience. For some practitioners, it may even be seen as a kind of theological statement: "There is no God; evolution explains everything in biology; therefore, human behaviors must be totally explained by evolutionary principles as well."

There is also the problem that many evolutionary psychology hypotheses don't work for the evolutionary time span assumed. Modern human behaviors like advanced language, sophisticated toolmaking, and artistic creativity have probably been around for between fifty thousand and one hundred thousand years (based on evidence from paleoanthropology). This is an extremely short time span for any kind of Darwinian evolution, and there is virtually no way that these distinguishing traits of behaviorally modern humans could have evolved through evolutionary adaptations within the time frame available.

If humans are not totally explained by evolutionary principles, then what other explanations for human behavior and human nature could there be? There are of course theological answers to that question as well as naturalistic explanations involving emergent qualities of our complex brains, but it is clear that human beings exhibit behaviors not explainable by natural selection in its fundamental meaning.

## How Human Nature and Evolution Support the Existence of God

As I have stated, Darwin's theory of evolution by natural selection is a powerful explanatory framework for a great deal of biology, though it cannot be successfully applied to all questions about life (for example, the origin of life, as we saw in the previous chapter). I believe that evolution is not antithetical to a belief in divine creation.[8] I want to go further here and propose that evolutionary mechanisms buttress the Christian view of a loving God who created the world and all that is in it, including humankind. I believe evolution actually *affirms* the existence of divinity.

How can I defend such an idea, which a large number of Christians would find absurd and antibiblical?

I believe an argument can be made that according to evolutionary theory, human beings should not behave the way we do. I think it is clear that adaptation and evolutionary selective advantage do not explain a great many human behaviors, such as altruism not directed toward kin, thoughtful reflection, creativity, humor, art, and so on.

Evolution allows us to frame the question of human behavior in terms of scientific knowledge about what makes any species what it is. We know what makes an earthworm an earthworm. It has earthworm DNA, whose genes and regulatory elements are the result of tuning by mutation and selection in a specific environment, which the earthworm itself shapes. Similar descriptions work well for most organisms. When it comes to humans, though, these descriptions only work to a point. Our bodies have clearly been shaped by evolution in the same way, but there is so much of our behavior that the DNA-and-natural-selection story doesn't explain.

If we discard evolution, the question could be easier—we might think there was nothing special about the human ability to master the world and exhibit a level of agency and understanding beyond anything else found in nature. It's just the way humans were created. But if evolution, while true and useful in most cases, cannot explain why we fall in love or sing in harmony, we have good reason to look elsewhere. And I believe we need only look to the words "made in the image of God" to find the answer.

I recently heard an atheist claim that human belief systems involving gods are programmed in us because our brains are wired to seek out agents and protective figures all around us, and this is why so many people are believers. This argument was presented as a scientific case against religion.

The argument is of course not original and has been advanced for some time. Yes, humans, as members of a highly social and cooperative species, do tend to seek agents and understand things happening in the world around them as related to the intentions of other minds. But beyond that, this is yet another example of the "just so" approach to human behavior favored by evolutionary psychologists. And the fact that we humans look for agency everywhere says nothing about whether such agency exists. We also look for sweet foods (because of the advantage in eating as many calories as possible in an environment where food is scarce), and sweet foods *do* exist. I would submit that this argument is very poor science indeed, and probably not worthy of the name.

As a counterargument, I can report on a recent experience of going to a concert by an African gospel choir. I am a trained musician, and I understand harmony, complex rhythms, chord structure, and so on. But the music I heard was beyond any of my previous experiences. The overall effect on me was one of pure joy and appreciation of a divine beauty.

A careful analysis of the melodic, harmonic, and rhythmic complexities of the music of the choir will almost certainly reveal no evidence of improving reproductive success of either singers or listeners, compared to such things as being large (for men) or shapely (for women). Of course, the same is true for all music—classical symphonies, jazz, and other styles of music with which I am more familiar.

The scientific logic behind the evolutionary approach to behavior applies quite well to birds, squirrels, and lions, but it really doesn't work for humans. There might have been a time when human beings lived in ways that were not very pleasant for most of the individuals but did optimize the population size by ensuring that maximally fit individuals mated the most and had the most optimally fit offspring. In many mammals (including some of our close ape relatives) the largest, strongest male would have a group of females to impregnate. At the same time, some of the females would be attracted to lone males from outside the group. This allowed for the physical fitness of the alpha male to be passed on, while at the same time providing for some degree of outbreeding, necessary for genetically healthy populations. It's not that we don't see traces of this behavior among people as well—we do. Women can be attracted to established, strong, "alpha" males and also to the "mysterious stranger" from outside the clan. But that is not how the majority of people live in modern times, or for the past ten thousand years, at least.

There is another way to answer the questions about why human beings do what we do. It starts with acknowledging the reality of one of our most important features: human consciousness. After all, it is the conscious mind of a person that is the ultimate source of all the behavior, emotions, and creativity that we have been discussing. That raises the obvious but extremely difficult-to-answer questions: What is consciousness, and where does it come from?

## Human Consciousness

Questions concerning human consciousness, or the nature of the mind, were long considered outside the realm of science—they were the province of theologians and philosophers. Concerning the mystery of human consciousness, Thomas Huxley wrote in the 1860s, "How it is that anything so remarkable as a state of consciousness comes about as the result of irritating nervous tissue, is just as unaccountable as the appearance of the Djin when Aladdin rubbed his lamp."[9]

I have always been intrigued by the subject of human consciousness. In chapter 7 of my book *The Works of His Hands*, I describe some of my ideas:

> Before I found God, while I was searching for an understanding of human nature, I liked the idea that consciousness might be something that emerges from the concentrated complexity of neural interactions in human

> brains. I thought of consciousness as an emergent phenomenon of complex life that cannot be entirely explained by the laws of biology—in the same way that life is an emergent phenomenon of complex chemistry, but it cannot be entirely described or explained by the laws of chemistry.[10]

I did not, in that earlier book, spend much time discussing the essence of human consciousness or any of the alternative nonreductionist or nonmaterialist theories of what it is and how it works. While the topic is huge, I will cover some of those things now. Starting with the connections between consciousness—which could be considered synonymous with the mind—and the brain, I will ask, as I did for human behavior, whether human consciousness has been demonstrated to be an outcome of evolution by natural selection.

### *The Mind-Body Problem*

One of the most difficult problems in philosophy and natural science has been the mind-brain or mind-body problem. This is the essence of the philosophical use of the term *dualism*. Despite an enormous amount of work, it has not been possible to bridge the gap between the limited phenomena we understand through the science of neurobiology and what we humans know and perceive as ourselves, whether this is called the mind, the soul, or the ego. Some have held that the problem of consciousness is not resolvable scientifically.

A large volume of written material has been devoted to a description of consciousness—its limits, its pathologies, and its overall nature—in an effort to understand its origins. But this is probably a hopeless task. A full description of everything we know about consciousness has not taken us much closer to understanding where it comes from, other than being a property of certain brains. Studies of neural connections, pathways, interactions, and anatomies have not really answered the basic questions: How does consciousness arise? Where does it come from? And what exactly is it?

We all know we have it, and we sense that some animals have it to a lesser extent. We know that it is involved with our large and intelligent brains but that it is not the same thing as intelligence. One way to sum up all the conundrums regarding the meaning and nature of consciousness could be to ask why we feel that we are conscious.

The answers that have been provided range from the denial of the very existence of the problem by some behavioral psychologists to attempts by physicists to apply quantum theory[11] or other known laws of science to the issue of consciousness.

I think one clue to consciousness lies in the very human concept of a story. Humans tell stories. Perhaps that is the main thing we do. We tell our own life stories and those of others. When we relax, we immerse ourselves in stories (films, sports, music, books, the internet—ultimately, it's all stories). So what do stories have to do with consciousness? Perhaps there is a good definition of human consciousness in the ability to organize experience into a story. There is no question that certain animals (among them dogs, dolphins, elephants, and other primates) can communicate in some form of rudimentary language. They can ask for things like food or attention, express emotions, react to events, and even relay information about something happening in the moment—the presence of danger or a food source. But it's hard to imagine any of them telling a story. As a writer, I love the idea of storytelling (as in this book) as a correlate of human consciousness. But while that might be true, it still doesn't get us very far in understanding where it comes from.

The standard scientific paradigm for how we can fathom anything has long been the reductionist approach: find the smallest bits, describe them in detail, and use those descriptions as bricks to rebuild the whole. Accordingly, the structure and function of brain components, including neurons, groups of brain cells, and brain regions, have been studied extensively in the hope that from understanding the physiological workings of neural tissue at this level, the answers to questions about consciousness would appear. This has not happened. And even the more innovative approaches, like invoking fundamental physical principles to explain consciousness, still assume that the answer will come through a bottom-up, reductionist methodology, such as through quantum state changes in individual neurons.

We also know very little about how human consciousness got to be the way it is, qualitatively different (as far as we can tell) from that of any other living creature. Over the years, I have seen a consensus of sorts emerge that holds that intelligence is a survival trait and can be explained by natural selection, but consciousness is not. No one has been able to explain how being conscious could provide a selective advantage to an individual or group compared to an intelligent but nonconscious being. If anything, it might be just the opposite. Unconscious robots perform quite well, after all.

### *A Thought Experiment on Consciousness and Evolution*

Could the origins of human consciousness be related to the evolution of intelligence, the use of language, and other unique or more highly developed attributes of human biology?

Intelligence is fairly easy to understand in terms of selective advantage. Many species have probably advanced in intelligence compared to earlier forms through the action of natural selection. Wolves and their close relatives, dogs, have evolved a fairly high level of intelligence, given their brain size. But consciousness is not intelligence. Computers are, at least by some definitions, intelligent, but they are not (at least not yet) conscious.

But if intelligence is subject to evolutionary forces, why not consciousness? Let's do a thought experiment. Imagine an early human, a member of our own *Homo sapiens* species, sitting on a plain in East Africa about 150,000 years ago. He has the same brain we have. He is smart and can make tools and use his wits to survive. And he has some degree of consciousness but not a lot. His tribe is just like him. They might even be able to speak, or at least communicate at a functional level. "You go there, wait for herd. I signal, you throw spear."

Next to them lives another tribe with the same level of intelligence, but by some unknown process, they have achieved the modern level of human consciousness. The members of this tribe, unlike their neighbors, feel things a lot more. They are emotional. They ask a lot of questions. They paint pictures on walls. They bury their dead and make beautiful objects. They sing and dance and derive pleasure from love. They are *us*.

Now, here is the question: Which tribe is more "fit" in evolutionary terms? Has the acquisition of consciousness contributed to any improvement of fitness for the "conscious" tribe? Let's imagine a herd of buffalo passing between the two tribes. The less conscious group gets up, the leader gives a few routine orders, and the hunt begins and is successful. On the other hand, over at the conscious tribe, when the leader says, "Hey, guys, I hear the buffalo moving," one of the hunters might say, "Not today! I am in a bad mood." Another might wonder, *What is the point of killing more game when we are all going to die anyway?* There might even be a few folks who think killing animals, even for survival, is "immoral," a word the nonconscious tribe could never understand. While this scenario is admittedly more tongue in cheek than scientific (you may even call it a "just so" story!), I believe the point is made.

The point is that consciousness does not really confer any selective advantage that would have meaning in the race for survival. "We cannot at present solve the 'hard problem' of how natural selection was able to produce consciousness, that is, answer the question of how consciousness emerged in the evolution of some animals."[12]

If it is true that consciousness does not confer any selective advantage, then we have a problem explaining its origin, given that its presence is universal in the human population. There is no reason—in fact, it should be impossible—for consciousness to have spread completely in the human population unless, like all other traits, it provides a selective advantage.

### *Consciousness, Codes, and Chemicals*

Software engineer Perry Marshall, the author of *Evolution 2.0*,[13] has come up with a creative way of thinking about the origins of consciousness. As he puts it,[14] in the general doctrine of biology, chemicals gave rise to code, which led to the beginnings of consciousness. In other words, abiogenesis began with chemicals that somehow led to the formation of the genetic code. And the genetic code, along with the rest of biochemistry, eventually produced the right proteins and biological structures—such as nerves, brains, and complex neural networks—that resulted in consciousness. But then Marshall points out that we have no idea how the genetic code could have been formed by chemicals simply doing chemistry. Not to mention that we still have no clue as to the mechanisms by which consciousness arises from the complexity of neurons and their connections.

On the other hand, we *do* know in great detail how conscious beings can produce codes (after all, that is Marshall's job), and once we have a code like the genetic code, we know precisely how such a code can produce a host of biochemicals, like proteins. The standard biological dogma that chemicals make codes make consciousness should be replaced by the insight that the reverse pathway is correct: Consciousness makes codes makes chemicals.

The implications are clear: Consciousness came first, before the beginning of life as we know it. This is of course not a new idea among philosophers and theologians, but Marshall provides a fresh statement from his computational perspective that ultimately buttresses theism. This view is growing among thinkers about the nature of reality and may soon become a standard argument for the existence of a creator God.

### *Consciousness and the Human Soul*

Humans are qualitatively different from other life-forms in how we act, reflect, create, communicate, and understand the world. The overwhelming evidence for how far ahead we are in some of our abilities is hard to deny, but some do so out of strong adherence to an ideological position that is not based on science, although it often pretends to be. The questions we are then faced with are the very questions that modern anti-humanists (for want of a better word) will claim are irrelevant. For me, however, they are quite salient. What triggered the ability of humans to be the unique way we are? What changed in *H. sapiens* (or perhaps in an earlier member of the genus *Homo*) that led to the way all humans are now?

Was it a spontaneous connection of neurons that allowed for a higher level of consciousness to arise and spread throughout the species? Was there an evolutionary pathway involving how we walk, use our hands, and give birth that gave rise to the intricacies of human language? Was it a mutation that led to an ability to recognize ourselves, to wonder at our mortality, to give us a sense of self, that no other creature possesses? Or was it an act of God to instill a soul into the starving, dying remnant of a desperate primate species on the brink of extinction?

What is a soul? A soul is something beyond intelligence; it is something that allows us to go way beyond the success of any creature that has ever come before. Our souls enable us to produce wonderful people with divine gifts, like the people who painted the pictures in the caves of France or discovered agriculture and built the first cities. The scientists, the musicians, all the geniuses, both known and unknown, and all the ordinary mortal folks, who no matter their station in life, are endowed with abilities and talents and insights that are truly remarkable from a biological standpoint—are the products of humans with souls.

The soul of each human is a divine gift from the Almighty. I know nothing in science that contradicts this, and I know nothing in science that can give a better, simpler, or more encompassing explanation of what we know it means to be a human being.

But isn't "soul" just another name for our conscious mind? Not exactly. A soul is a religious concept and is not addressed by science. The way I see it, God could have chosen to give souls to other species instead of us. In that case we would just be a regular species of primates, eating worms and berries and acting much like the others, while perhaps the dolphins or the wolves would be joining internet chat rooms, falling in love, writing poetry, and dreaming of all manner of things.

But God's gift was to humankind, allowing us to flourish as a totally new kind of creature, one that is not satisfied merely with adequate food, water, and shelter. With our souls comes the thirst for knowledge, an understanding and appreciation for beauty, art, and music, and the creation of beautiful things for their own sake, beyond survival-related purposes.

A recent book by Yale professor Samuel Wilkinson titled *Purpose: What Evolution and Human Nature Imply about the Meaning of Our Existence* focuses largely on the human psychological need for a sense of purpose. He quotes the great science writer Lewis Thomas (author of *The Lives of a Cell*) as saying, "I cannot make my peace with the randomness doctrine; I cannot abide the notion of purposelessness and blind chance in nature. And yet I do not know what to put in its place for the quieting of my mind."[15] Dr. Thomas was not a theist.

Wilkinson writes, "There is emerging evidence that brain systems (and the behaviors they give rise to) are not deterministic. . . . This has even been seen in comparatively simple brains, such as those of fruit flies, leeches, and microscopic roundworms."[16]

### *Consciousness and Mind*

This brief discussion of our God-given human soul does not, of course, address any of the scientific questions about the nature and origin of consciousness. If we think of human consciousness as being connected in some way to the human soul, then this is tantamount to admitting that both consciousness and by implication the human soul stand apart from the theory of evolution by natural selection.

The standard scientific approach to consciousness is the physicalist or monistic idea that all aspects of what we call "mind"—which include cognition, memory, emotions, thoughts, and so on—are fundamentally causally connected to brain neurophysiology. One theory holds that memory states are stored in changes in the synapses (connections between neural cells), spread in many regions of the brain.

While there is a great deal of evidence for this view, there is also lots of evidence that such a physicalist basis for mind does not sufficiently explain the experimental and observational data. Physicist Marco Masi has outlined much of this contrary evidence in his article "An Evidence-Based Critical Review of the Mind-Brain Identity Theory."

There are many problems with the idea that the mind is entirely a property of the brain. Masi acknowledges the evidence that physical brain states, including those induced by drugs or injuries, have direct effects on consciousness. But he points out that this evidence of correlation is not logical proof of causation. Indeed, there are many examples indicating that conscious experience, including emotions, cognition, and awareness, may not be entirely or sufficiently explained by physical brain states. Masi writes, "To date, there is no evidence, not even indirect or circumstantial, of a single brain region, area, organ, anatomical feature, or modern Cartesian pineal gland that takes charge of this mysterious job of 'producing' or 'generating' consciousness."[17]

In a *Psychology Today* article titled "Why Science Will Never Explain Consciousness: Explaining Consciousness in Physical Terms Is

Conceptually Impossible," Sharon Hewitt Rawlette writes, "There's an essential logical link missing between the kinds of processes investigated by physical science and the conscious experience that sometimes accompanies them. And this makes a physical explanation of consciousness impossible."[18]

Modern neuroscience has more or less abandoned the idea that the brain is entirely compartmentalized, with each region or section responsible for specialized functions. This view has been replaced, based on experimental evidence as well as ideas from network science, by a view in which various brain regions function holistically, in a complex system of interconnectedness and interdependence.

So if there is no strong evidence for the materialist view of complete brain-mind identity, how do we explain the phenomenon of conscious experience (other than as an illusion, as suggested by Daniel Dennett)? Masi suggests an alternative model based on classical dualism that supposes a "transmissive" function of the brain, as proposed by philosopher William James. In other words, we are not actually the producers of our thoughts, feelings, and ideas but are transmitting them from the outside, with our brains acting as filters of an entirely different form of awareness.

James used the analogy of a prism to explain what he meant. A prism transmits and alters visible light into a spectrum of colors as white light passes through it. It is not the source of the light. When a prism is broken, it no longer produces the beautiful spectrum it did before, but this does not prove that the prism was the source of the light before it broke. In the same way, a broken brain may fail to transmit consciousness rather than fail to produce it, as the body-mind theory would have it. This view is consonant with the idea that consciousness may exist in the absence of brains or any kind of neural system. As we saw in chapter 4, forms of cognition can be found in organisms and networks of organisms such as plants, bacteria, and so on without any neural structures at all—although it must be said that there is no evidence that such organisms are conscious in the same way we are.

While there is only circumstantial and weak scientific evidence to support the idea of the brain as a transmitter rather than the producer of conscious thought and personality, I find it theologically and philosophically appealing. In this view, the ultimate source of the "mind material" we all transmit could be God. We cannot even guess at how such a thing might work, but the idea that we "think God's thoughts after him" is not a new one.

And of course, if each of us is a prism transmitting the light of God's consciousness, we each do it in our own way, in part because of our different brain chemistries but also because of the ineffable personal ownership of different individual souls. In this way, we might think of the brain as the receiver and transmitter of cognition, and our souls as the translator of this information into emotions, thoughts, and flashes of insight, even of genius.

Many writers have remarked that sometimes the words simply seem to write themselves, and I have experienced the same thing from time to time, including while writing the previous paragraph. We might ask, what is it (or better, *Who* is it) that is doing this ghost-writing for us? My answer has always been clear. It is the blessed visit of the Holy Ghost.

## Consciousness and Computability

The fact that it is so difficult to explain consciousness in scientific terms raises the possibility that it might not be possible to do so. Brilliant physicist and Nobel laureate Sir Roger Penrose argues, using Gödel's theorem, that consciousness cannot be computational. In other words, there is more to human consciousness than can currently be explained scientifically. The reasoning is not simple to follow, but I will do my best to present it.

Gödel's theorems are among the most important (but also difficult to understand) breakthroughs in modern mathematics. They are related to the vexing problem of self-referencing language.[19]

As an example of the problem, take the following statement: "This sentence is false." If the statement is true, then it's false, and if it's false, then it's true. At first sight this might appear to be just a silly trick of self-referential language, but the problems associated with self-reference are not at all trivial. They formed the basis of Gödel's deep investigation into the theory of arithmetic systems.

Gödel's two incompleteness theorems say that any logical system contains either contradictory statements or statements that cannot be proven. A consistent logical system is composed of a set of axioms that allow you to do arithmetic according to rules related to the axiomatic statements. Gödel found that in any such system there must be at least one statement that is unprovable using just that system's axioms.

Gödel was able to prove these theorems in a way that is far beyond my understanding. Penrose says that Gödel's theorem shows that consciousness cannot be fully explained by any kind of numerical computational process because in order to determine whether the computation done within an axiomatic system is actually telling us the truth—rather than merely computing correctly according to the rules of the system—we have to step outside the system. We need "understanding," by which he means consciousness or a feature of consciousness. Therefore, consciousness is not computable.[20]

The implications of this idea are quite staggering. For one thing, it suggests that while AI might become very intelligent, it could never become fully conscious in the human sense. It also raises the question of how consciousness could arise in human (and perhaps other) brains if things like computational complexity cannot explain it. Indeed, it makes the task of even defining consciousness in a formal scientific sense extremely difficult, if not impossible.

## The Most Important Thing of All

So far, in this entire chapter on the nature of human behavior, consciousness, and souls, I have only mentioned in passing what I feel is the most important part of being human: love. Some will of course say, as they do for language, art, and creativity, that love is an evolved trait arising from mammalian reproductive biology. There are species of animals and birds that do appear to exhibit some aspects of love, including mating for life, shared care of offspring, and even mourning for the dead. So, again, we can find roots of humanity in earlier life-forms. But—also again—those roots do not come close to a full understanding or description of human love.

The subject is vast. It is the major theme of the great majority of stories we tell each other, from sitting around the campfire to movie blockbusters to most novels we've read. But I am not speaking here only of romantic love. Love is the most important action of the Christian faith: "Love your enemies," Jesus tells us, adding an unexpected plot twist to the older "love your neighbor" teaching (Matthew 5:43-44). "Now these three remain: faith, hope and love. But the greatest of these is love," the apostle Paul echoes (1 Corinthians 13:13).

The love between friends and relations, parents and children, and even between strangers who meet and perhaps discover mutual interests and ideas is the feeling that binds us. Love defeats loneliness, despair, fear, and anxiety. Love brings us hope and peace. Love is available for all humans, not only those in the process of sexual reproduction. We might ask, "What is love?" And—this should come as no surprise at this point—we do not have a good scientific answer. Again, there are naturalistic answers like "Love is a result of hormonal effects on neurochemistry." As we've seen with some of the answers above on consciousness and behavior, this is a sciency-sounding explanation with at best limited value.

Philosopher of religion Paul M. Gould has a chapter in his book *A Good and True Story*[21] about love, which he summarizes in an article in *The Worldview Bulletin Newsletter*. He writes, "The five things I notice when I gaze at love are that it is *multi-dimensional*, *complex*, *deep and enduring*, *active and passive*, and *valuable*." After describing each of those characteristics, he closes by telling us that "a world without love is a harsh and cold world. Our world is better because of love. . . . And it is love that I am called to follow as I seek to love God and others."[22]

I fully agree. For me, if there were nothing else about human nature that points so strongly to our divine creation, the existence of love alone, in all its holy glory, would be enough. "For God so loved the world that he gave his one and only Son, that whoever believes in him shall not perish but have eternal life" (John 3:16).

## Conclusion

I doubt that anyone who thinks that humans are nothing particularly special or valuable has come this far in the book, so I feel safe in assuming that you agree with my most basic premise (one I have held all of my life) that there is something about humans that is clearly transcendent to the world of ordinary biology—which itself, I believe, is transcendent to the world of molecules, as I argue in much of the rest of this book. I have searched for the source of human transcendence for a very long time, found and rejected many possibilities, and now know, as surely as I know that Jesus Christ is my hope and redemption, that the source is a wise, powerful, and loving God.

CHAPTER 7

# Divine Design: The Beginning

BIOLOGY PROCLAIMS THE GLORY OF GOD. Even though our current understanding of biology is incomplete (as is all science), rather than using the idea of God to fill gaps of what we don't understand, we should fill those gaps with new knowledge and theories that will point to God as the Creator in a similar way that Big Bang cosmology and the fine-tuning of the universe have pointed to God in other branches of science.

If you've read my previous books or my blog, you know that I sometimes attempt to make a point with a story or fable. In this book I have treated you to only one of those. I think you deserve one more as a reward for getting this far. Here it is.

On a beautiful day in heaven (where all the days are beautiful), Werner Heisenberg met up with his good friend Louis Pasteur, and as was their custom, they started on a walk through the heavenly woodlands. As usual, they spent most of the time talking.

"Louis," Heisenberg began, "have you heard of this Garte fellow? An American, I believe."

"Who?" Pasteur replied.

"Garte. A biochemist. Not terribly brilliant, but he has published a lot of papers and written a few books. The most recent one is interesting."

Pasteur shook his head. He had always kept up with the development of biological research in the world but had become a bit jaded by the lack of much progress in things he was most interested in since the feverish activity of the 1950s and 1960s.

"Does he mention us in these books of his?" he asked.

"Indeed, he does, at least in passing."

"So, what's his point?"

"He seems to be predicting a new phase in biology, much like the grand old days we physicists had in the 1920s."

"Indeed?"

One of the main topics of conversation between these departed souls of Christian scientists was the contrast and comparison of physics and biology as sciences. Pasteur well remembered watching, from above, as modern physics unfolded. He was of course never envious because there is no envy in heaven.

"Yes," Heisenberg said. "He seems to be saying that biology is on the verge of some major breakthroughs related to many stubborn problems."

"Interesting. And what, pray tell, does this Gord fellow presume is the answer to these questions?"

"It's Garte—with an *e* at the end. Of course he has no idea, but he is proposing that the lack of formal laws in biological science, as compared to physics—"

"Now, Werner, please don't start that again."

"I am only quoting him, and he is a biologist! Anyway, he thinks that biology needs to incorporate teleology, agency, cognition, and design into these new putative theories."

"Does he? How brilliant of him. What next? Understanding that the word *function* as used in biology is never found anywhere else?" Pasteur tended to use a slightly ironic tone when speaking of modern scientists and their insurmountable philosophical ignorance. "And how does he propose to insert such things as purpose into a mathematical law of anything in biology?"

"He doesn't. He admits that he is not smart enough to do that, and I take him at his word. But he seems to be predicting that somebody will."

"Hmph."

"He does get one thing right, though, although it is not entirely original. He makes a strong claim that high-fidelity replication is a prerequisite for evolution."

"Oh, does he? Has he heard of Eigen?"

Heisenberg sighed. "Louis, he is not writing to scientists but to ordinary people, trying to explain to them how biology proclaims the glory of God."

At this point in their walk, they were joined by another physicist, Templeton Prize winner John Barrow, and a young woman I'll call Brittany Lee. The latter (entirely fictional) was one of the smartest people to ever live and had, as a graduate student at MIT, discovered the mechanism for the origin of the genetic code. She had typed it all up on her computer but then died while crossing Memorial Drive in an accident caused by the combination of bad signage and a confused parent attending the undergraduates' family weekend. Her computer was also destroyed, so no one on earth had ever heard of her discovery, but in heaven she was granted her due (like countless others).

It turned out that Barrow and Lee had both heard of Garte. Barrow had actually met him at a conference, and they had spoken briefly, though not about science. Brittany Lee had read Garte's first book about coming to faith in Jesus as an atheist scientist.

"Brittany," Heisenberg asked her, "how do you think this man

Garte will feel when he gets here and finds out the ultimate truth about life and its origins and everything he wanted to know about human nature?"

"He will probably react as I did—as we *all* did: with astonishment, humility, and worship of the Creator. I will never stop marveling at the works of his hands."

"Yes," Pasteur said. "He will find out that he was wrong about so much but in ways he could never have guessed. Just as it happened to you, Werner."

"Well," Barrow put in the last words on the subject. "I hope his book does well. It might be useful to many people, even if it's only slightly true."

At that point the conversation turned to the subject of artificial intelligence, which was consuming their colleagues on earth with wonder and worry, and they yet again had a good laugh at how little the humans on earth understood.

## A Summation

This book has been about what I hope and pray will become the beginning of something in the realm of the science of life. It might seem to you, as it has to me for some time, that this dream of a new reconciliation of biological science with Christian faith is only that—a dream. If that were true, if everything I have written and proposed in this book were nothing more than my own thoughts and hopes, I would not have written it. But every week I am reinforced in my sense of excitement that we are witnessing the beginning of a sea change in the world of biology, as demonstrated by the growing wave of scientific publications like those highlighted in chapter 2.

We have seen that whether evolution is true or false is not worth fighting about. We have seen that the evolutionary process is not a naturalistic alternative to God's creation of life, evidence for atheism, or in opposition to biblical truth. We have highlighted the far more important role of high-accuracy self-replication of cells, Darwin's strong principle of heredity by which all of life exercises reproduction "according to their kind." And we have seen that while we understand the amazingly complex biochemistry of that process, we cannot explain how it could have come into being based on only the laws of chemistry.

We have also dived into the murky pool of life's origins, seen the enormous difficulty of finding a theory to explain it, and come up gasping for air as we search for some degree of clarity about this great mystery. But there is light at the end of the tunnel. We are living in a moment of scientific breakthroughs in biology, where bold seekers, both those of Christian faith and others, are finding new pathways for pushing back against some of the dogmas that have restrained progress. Discoveries of the true importance of purpose, agency, and cognition in all of life from its beginning give us a strong sense of hope that we can indeed continue to move forward to the truth behind the miraculous phenomenon of being alive.

When we looked at the question of consciousness, we saw that being human is not just about being alive with slightly better faculties than the rest of the biological world. It's about a baby's first laugh, a deaf composer's symphonic masterpiece, the immeasurable joy of love, the finding of a beautiful and perfect equation, and the brotherhood of humanity.

Is all of this evidence for a creator God, for design from an intelligent source? I believe it is. But we need to frame the question carefully. There will never be *scientific* proof of God because divinity is not approachable by the scientific method, and science doesn't deal with proof. Science and faith are not in conflict but are two different and harmonious ways to find the same beautiful truth. We don't use science to prove our faith, and we don't use faith to answer scientific questions. We use both in different ways, to find the same truth about God and his creation.

## Divine Design in Life

Some of the arguments and reasoning you have read in this book might strike you as reminiscent of intelligent design. Of course, it's impossible to be a biologist (or at least an honest one) and deny the reality of design in biology. The hard part is understanding the source of this design. As I have shown, while evolution explains a great deal of biological design, it cannot explain *all* of it because life that is capable of accurate self-replication has to exist before it can be further designed by evolution. This certainly does not mean that design proves evolution to be wrong, as many in the intelligent design movement propose. It simply means that evolution is not enough, by itself, to explain *all* of biological design. The process of evolution, which follows naturally from high-accuracy self-replication, is only one part of the overall picture of design in life, what I call divine design.

So yes, I believe that life was designed, but also that calling that design "intelligent" is wrong because that implies the process is something taking place at our own, human level. That cannot be correct—after all, no human has ever come up with the design to make a living being from scratch, let alone been able to produce one using all the knowledge we have about the materials and mechanisms that make creatures alive. Furthermore, the proponents of intelligent design often frame their arguments for evidence for God in terms of the gaps and mysteries within our knowledge of biology (and the rest of nature). In contrast, the divine design concept is based on the new knowledge that fills those gaps.

## God of the Gaps

While most popular with atheists, the phrase "God of the gaps" is credited to nineteenth-century theologian Henry Drummond, who noted that "there are reverent minds who ceaselessly scan the fields of Nature and the books of Science in search of gaps—gaps which they will fill up with God."[1] He was the one who first pointed out the danger of looking for God in what science couldn't yet explain. What happens then, he asked, when science does fill the gap? But he went further than that—he argued that even if science can never answer some "how" question, we shouldn't see God only "in the disorders of the world"[2] but ever present in his creation, as Judeo-Christian theology in fact holds.

Why, then, is this concept so often used against Christians? It is certainly true that ancient humans, who had no scientific knowledge of how the world worked, inserted mythical gods and other unseen forces to explain things like thunder, illness, rain, and so on. But such thinking has been thoroughly discredited, and many Christian thinkers have long agreed with Drummond that using God to explain any natural phenomenon that has not been explained by a naturalistic mechanism runs the risk of making God too small.

Modern religious thinking largely separates religious claims and explanations of natural mechanisms. In the Western world, most Christians would not go to their priest or minister with questions about diabetes or farming techniques. Christian geologists don't point to God as the cause of earthquakes—they talk about tectonic plates. Christian biologists know that diseases are caused by microorganisms. And ever since Newton, Christian physicists have talked about the force of gravity keeping the planets in their orbits, not about God or angels playing heavenly ball games.

## The Alternative to Gaps

One of the keys to understanding the basic harmony and compatibility between science and faith is to forget about the gaps. The whole universe is God's, and we can use *what we do know* about scientific explanations for natural mechanisms, rather than what we don't know in order to find evidence for and pointers to God. We just need to remember that scientific explanations for natural phenomena are not the end of our understanding or our curiosity about the world.

When physicists discovered the existence of discrete lines in electromagnetic emission spectra from distant stars, a large gap was closed in our understanding of the mechanisms by which light behaves. But when Max Planck and Niels Bohr were asked what the mysterious reality revealed by quantum mechanics could mean, they answered that this was not of their concern. The science was correct because it worked, and it was up to others to work out the philosophical stuff. Many scientists have seen the hand of God in the revelations of our new knowledge of quantum theory and other areas of modern physics.[3]

In biology, the huge gaps in our knowledge of how inheritance works were beautifully filled by the discoveries of the structure of DNA and the ribosomal machinery. And yes, these discoveries were consistent with chemistry and biochemistry. But . . . not completely. Once we learned that DNA operates as a code, a new question arose. How could a code have arisen from purely chemical processes? This question remains unanswered. These examples from physics and biology illustrate that as we probe deeper into the nature of reality, we find that closing a gap of knowledge with scientific information often opens a new gap whose solutions may not be subject to purely scientific investigation.

## Emergence and Divine Law

In my most recent book, *Science and Faith in Harmony*,[4] I discuss the subject of emergence and how it might relate to Christian theology. Extreme complexity can result in emergent phenomena that are not entirely explained by the laws that govern the systems from which they emerged. Thus, atoms and molecules emerge from the laws of physics, but their behavior requires new laws specific to chemistry, like molecular orbital theory and bonding theory. Life is an emergent phenomenon of complex chemistry, but it is not entirely described or explained by the laws of chemistry. New laws of biology, involving enzyme kinetics and genetics, are required to make sense of how life works.

Similarly, consciousness is an emergent phenomenon of complex life that is not entirely explained by the laws of biology. New concepts and methods of psychology and neuroscience are required to understand how minds and brains work—even if they fall short of explaining the human soul.

If we go backward in time, we can perhaps imagine getting to a point where we see the universe of matter and energy as an emergent phenomenon of God, not entirely explained by the "laws of God." New laws of physics are required to understand the emergent universe.

What do I mean by the "laws of God"? Here we have stepped outside of science, of course, and into speculative theology. We assume that God existed before and outside of our universe of matter and energy and is the source of all emergences to higher levels of complexity. While we cannot know the "laws of God"—God's ultimate nature—we are given some clues in his Word that they might include something like the following:

- *The law of love.* God's creation is made from love, and all those who bear the image of God must love him and his creation, including other creatures.
- *The law of emergence.* All phenomena contain the seeds of the emergence of new phenomena as they grow in complexity and goodness.
- *The law of free will.* God created humans with the power to act on their own, for good or for evil, to live meaningful lives or to violate God's law of love.

In some universes, perhaps heaven being one of them, it might be possible to violate the laws of nature such as the second law of thermodynamics, or perhaps there are no such laws.[5] In such a universe, perhaps everything that happens is determined solely by the will of God. There, life could exist without death, sin, or suffering. But in *our* universe—in our world that operates according to laws of nature—life is not possible without death, and the complex, miraculous life that we have is not possible without suffering.

Why did God choose to create us in a universe governed by natural law along with his will, rather than directly placing us into a heavenly universe, where his will alone guides all things? I don't know the answer, and I'm sure we will not know the answer in this life.

We cannot begin to fathom the true nature of God, or the ultimate truth about the way he has brought about this world. The biblical book of Job makes this quite clear when God answers Job with the words "Where were you?" (Job 38:4). Job's response, acknowledging his own humble ignorance, is a model for us as well when faced with the overwhelming evidence we now have about the unending complexity and majesty of nature.

## Evolutionary Philosophy

As we have seen, evolutionary theory does not address, and was never meant to address, underlying philosophical mysteries about life—questions like Why is there life? Is there a purpose to life and, in particular, to conscious, sentient life? As discussed in chapter 2, answering such questions by reference to evolutionary mechanisms is not good science, and it makes for bad philosophy as well.

Unfortunately, despite this, some religious skeptics have placed evolution on a pedestal, a place where it does not belong. Evolution is not an all-encompassing principle, the answer to "life, the universe, and everything." No matter what we learn about the remaining scientific mysteries of life, including its origin, I believe that the deeper questions of the meaning of life will remain to be answered only by theological and philosophical inquiry.

A thoughtful and beautiful account of how evolution can be seen as an instrument of God's design is given by Jim Stump in his recent book *The Sacred Chain*. Stump, a philosopher with a deep understanding of evolution and of Christian faith, shows both skeptics and evolution-wary Christian believers that the basic theory of how life evolved and gave rise to us is not in any way opposed to a biblical view of creation. One passage from this book makes the point:

> God didn't create things originally the way they were ultimately intended to be. . . . Instead, God took a long time, not being concerned about time and efficiency . . . and let evolution produce the right kinds of body types that could give rise to the right kinds of souls, ultimately allowing language to develop, along with the ability to see things *as* other things. . . . God created in a way that would bring us about but also so that an extraordinarily lavish variety of other life could exist.[6]

## The Theology of Evolution

Among my Christian friends, there are some who continue to believe, in some cases strongly, that the theory of evolution contradicts the biblical account of the creation of life. In response, I say that Genesis is not teaching us science but about God's purposes with creation. I also point out that, as I've shown in chapter 3, evolution is not the whole scientific story behind life, and the repeated phrase "according to their kind" in Genesis 1 could be seen as an excellent illustration of the idea that inheritance (or high-accuracy self-replication) is the foundation of all life. By stating that God created all the animals in the sea, sky, and earth according to their kinds, the Bible says nothing in contradiction to evolution. It even confirms it—in language that would have been understandable to its audience of ancient farmers. That all animals always and only give birth to their own kind is not only a basic fact of biology but also an absolute requirement for evolution, and a direct and powerful statement of the principle of inheritance.

The Bible, of course, doesn't mention common ancestry, phylogenetic trees, genetic mutations, or any of the other details of evolutionary mechanisms known to modern science. It is not a science book. But in its metaphorical language, Genesis is telling us the truth. Life, as created by God, is based on a phenomenon whose details are still far beyond our scientific comprehension: the ability of a collection of chemicals enclosed in a membrane to reproduce itself with high accuracy.

I believe a good deal of Christian resistance to evolution stems from scientifically unsupported claims by biologists that evolution is a "blind" process, reliant on random mutations and without any apparent direction or purpose. As mentioned in chapter 2, Stephen Jay Gould stated, based on this principle, that if we could wind the clock backward and start life from scratch, we would see an entirely different world of life, and probably no people. This conclusion has since been strongly challenged, if not disproven, by Simon Conway Morris's (and others') studies on convergence,[7] as discussed in chapter 2.

Some further thought about evolution should lead us to consider that if Gould (and many other biologists) were right and evolution is completely random, then the emergence of large creatures with brains and the cognitive capacity to worship God was fairly unexpected and unlikely. But if, on the contrary, the path of evolutionary change was in some way guided to lead to us, perhaps by as-yet-unknown God-given laws of biology, then again we find reason to say that our existence is a miracle of creation. Either way, if we want to see the work of God, we need simply look in the mirror.

We can argue that evolution is very much a part of God's plan for life. This would mean that Darwin's idea of evolution of species by natural selection, instead of being in conflict with Scripture, is in fact among its scientific supports. God's creation had in it exactly what was needed for life to evolve into creatures who would sing his praises, portray his glory, discover his laws, and write about his mysterious ways. God's will to have living animals capable and desirous of a relationship with him is, then, the Aristotelian "final cause" or teleology of evolution.

We could add that today, in the information age, we also know that allowing a system (such as a complex computer algorithm) to develop through a process of variation and trial and error (similar to the way natural selection works in Darwinian evolution) is more efficient and simpler than to create new, specific programs from scratch. In fact, in computer science such methods are called "evolutionary computation." It is therefore not at all surprising that God would have used this method rather than create millions of species separately to produce the final desired result—beings who are able and willing to recognize their creator and worship him.

As for the origin of life, there is so far no possible evolutionary theory that can account for it, as we have seen in chapter 5. If God had created all species *de novo*, then the origin of life would be no mystery—in fact, it wouldn't even be a question. It would mean all life began at the same time that God created the first of every given species, exactly as they are today. On the other hand, if evolution is true, then there had to be a beginning of life that was much simpler than what we now see all around us, and it is the origin of that simple cell, in the absence of evolutionary processes before evolution was possible, that science has not been able to explain. In this sense, then, it is the very reality of evolution that makes the origin of life a scientific conundrum—and a pointer to a creative divinity.

The Bible is the divine Word of God, but I think we are justified in believing that God expected every generation to interpret the

Word according to the age in which they live. Those who dispute this put a major limitation on the workings of God, as well as on us, the pinnacle of his creation, by implying that God assumed there would be no human progress with time, and that what rang true for us thousands of years ago would still ring true today. This is especially true for the nature of what we think of as "scientific truth." I see no contradictions between the holy, divine-inspired words of the Bible and what we have learned from scientific investigations into the nature of the universe and life on earth. One reason for this is the ever-changing nature of scientific knowledge.

## Scientific Truth Is Not a Constant

There is not a great deal of actual science in the Bible, and what there is seems to most modern people to be just plain wrong. Of course, it would be miraculous if the Bible writers had accurately portrayed the geology and geography of the earth or had understood the genetic and cellular basis of life. But isn't it reasonable to assume that such a miracle could have (should have) been accomplished by an all-powerful God?

I have often been asked why God didn't correct the scientific "errors" in Scripture as time went on. After all, God came to earth to live, teach, and work among us, and Jesus could have explained the truth of cosmology and biology to his disciples. And yet in all the New Testament, there is hardly a single word from Jesus, any of his disciples, or Paul in his many letters containing any scientific content. It is worth thinking about why that is—why Jesus quoted Scripture often in the Gospels but didn't seem interested in scientific truth.

What are we to think about this? Did God mean for us to remain scientifically ignorant and for people in the twenty-first century to follow the four-thousand-year-old "science" of ancient days, no matter how out of date it is? Clearly not. Even (most) biblical literalists know that the earth is not flat (as some poetic passages could be read to suggest) and there is no solid "firmament" in the sky.

The answer lies not with God but with us. When we talk about scientific truth, we need to ask ourselves which scientific truth we mean. If Jesus had decided to teach science to his followers, what science would he have taught them? The science of his day? That wouldn't have been such a great improvement over the Old Testament version. Should he have said, "Let me explain the way the natural world works" and then described quantum theory, the Big Bang and galactic expansion, the periodic table, the genetic code, and so on? In other words, modern science—the science of 2025?

But why should he have chosen 2025 rather than, say, 1875? People living in 1875 also thought their knowledge about the natural world was the ultimate scientific truth, but at that time there was no quantum theory, no genetic code, and the chemistry and astronomy they knew would not be quite right either, from our point of view. Of course, had he chosen the science of 1500, it would have been even less developed.

In other words, consider that it just might be possible—in fact, it is rather likely—that 2025 science is not as correct as the science of 2050 or 2500 will be. I think we can be sure that the "correct" science will change because it always has before. The science of today will very likely be seen as containing many errors by our descendants in the coming centuries.

The question of which science God "should have" instructed his scribes to put to paper is not one we can answer. It's something God knows because he is God, and he alone knows the final truth of how the universe and everything in it works. He created it all. If back in Jesus' day, or even today, he had made it known somehow that in fact the trunoligous wave particle controls all photonic collindations from dark matter into Bulisinski energetics, or that mega-crypto-nucleotide repeats regulate transmembrane activation of bridging peptides for epigenetic reversal of Stettmeyer fibrules, which of us would even understand his use of this common scientific parlance from the twenty-second century? (All of which I just invented, so please don't bother looking it up!) My point is that whatever future science will say, we wouldn't understand it if we read it in Scripture any more than Jesus' first-century audience would have.

It isn't God's will to tell us the science of how the world works. It's up to us to figure that out. We have his Word that tells us how we should live, how to love our neighbors, whom to worship, and how to be the best humans we can be. And we have his creation—his book of works—to learn about the natural world.

During the period that I was wavering between agnosticism and faith, I came across this famous verse: "For since the creation of the world God's invisible qualities—his eternal power and divine nature—have been clearly seen, being understood from what has been made, so that people are without excuse" (Romans 1:20). It struck me that Paul was telling us that nature ("what has been made") reveals the reality of God's existence and qualities. My first thought was of the creation of the universe ("the world"), but I was a full-fledged biologist, and I realized that the world of life is also "what has been made" and that perhaps I was indeed without excuse.

As a scientist, I am drawn to the notion of digging as deeply as possible into how everything happened, but I also know that using our science of today, we can only go so far, and at some point, the only way forward is to stop and pray our thanks for this wonderful world we were given and for the gift of love that surpasses all understanding.

## Science and the Bible

Genesis tells us that God is not the sun or the moon. The heavenly bodies are creations of God, not gods themselves, and the same is true for all of creation, for living beings as well as nonliving objects.

Humans are created in the image of God to be God's representatives on earth, not to be the gods' servants, as in many ancient Near East myths. The story also tells us that this creation did not happen all at once but in stages. We read about the creation of one thing after the other in order, for God's creation is ruled by order, not the chaos of ancient myths.

The existence of natural order is what makes our search for scientific truth possible, and I find the expression of this order in the symbolic and transcendent text of the Bible truly inspiring. It should make all persons of faith rejoice and kindle in us a desire to understand more about God's creation through science, knowing that it is another way of honoring God.

When people find what they believe are contradictions between the words of the Bible and modern biological science, they are interpreting the Bible in a particular way, which they call a "literal" or "plain" reading. However, many who espouse this method of biblical interpretation are often quite selective in applying it. An ironic example is the conclusion by some Christians that the world is six thousand years old, despite the fact that there is no *literal* number given for the age of the earth in any part of the Bible. Instead, this conclusion comes from an *interpretation* of the genealogies in Genesis, generally relying on Bishop Ussher's calculations in the seventeenth century.[8]

Furthermore, we know that God speaks to us in many ways, not only literally. Jesus Christ, God incarnate, taught his disciples in parables whose true meaning was never literal, and one can find obviously allegorical, metaphorical, and symbolic truths throughout Scripture. Finally, while there is no question for a Christian that the Bible was inspired by God, the hands that wrote the words down were human. And humans are limited in their understanding and knowledge.

What I find miraculous and joyful is that reading God-inspired words recorded so long ago can move us today to see ourselves and our neighbors as creatures made in the image of God and help us live meaningful lives through that knowledge, loving our neighbors and building the Kingdom. We just need to keep trusting in the Lord and be assured that by following his two books (God's Word and God's works), we will follow him to glory.

## The God of Wisdom

We have begun to find that many gaps in our scientific understanding of nature have been filled with answers that might have once been considered tinged with something other than "natural" or rational knowledge. If we were to try to explain the reality of quantum theory, general relativity, the Big Bang, or the details of cellular molecular biology to the scientists of 1890, it's quite likely some would comment that it all sounds irrational or even supernatural.

While scientists who are Christians believe that this world is a divine creation, we also respect and accept the scientific method and use our God-given abilities to find answers to questions about how (and sometimes why) God acted to create the world as it is. This is what led the earliest Christian natural philosophers like Bacon, Newton, Boyle, Copernicus, Galileo, and Pasteur to invent and perfect the scientific methodology used today. And when they used those methods of experimentation, observation, repetition, and control of variables to study the motion of the planets, the laws of chemistry, or the nature of the living world, they were praising God for his creation in a way that was as deep and as pure as a congregation singing a hymn and praying in a church service. This is why I say that science is distilled doxology.

Scientists of Christian faith used established scientific methodology to obtain knowledge, and by doing so found pointers—not to a God of the gaps but to the God of wisdom. That God is very real, and every new discovery is pointing to him. When we finally understand the mysteries of how an information-bearing molecule and the machinery needed for accurate cellular self-replication could emerge at the origin of life, we will have even more solid evidence for the God of wisdom.

## Final Thoughts

Richard Dawkins is almost right when he claims that "at bottom" the universe is pitiless and indifferent. In all the billions of galaxies, among the uncountable number of stars, planets, moons, and asteroids, nothing has any deeper meaning or purpose than to follow the laws of physics and chemistry.

But there is one exception. One place where meaning, purpose, agency, beauty, love, music, and goodness exist and thrive alongside the indifference of natural law. In this place, our planet earth, there is life—something novel, special, and beyond all the natural laws that govern everywhere else.

"Why?" we ask. "Why is there life, what is it, and how did it start?" We humans can and have learned a lot about life. We invented a science called biology. And in many cases, what we have learned only makes those enigmas we cannot resolve stand out more visibly.

We have seen that there is no clear reason that life should arise from a sea of chemicals. In fact, not only do we not know the answer to that question, we *do* know enough about chemistry and physics to be able to say, "There is no reason that life should exist." We can exclude any scientific reason for the existence of life based on all the science we know. We can see there is no reason for anything to be alive. Where does this leave us?

We could just leave it at that and conclude that there is no "why" about life; it just is. But perhaps the answer is not to be found in the light under the lamppost we have been staring at. The light we are using to try to find the answer is shining on a world of chemistry and physics that does not include the missing key. What we need to do is build a new light and use it to look more closely at life itself. When we do that, we see something that does not exist in the rest of reality—a strong sense of purpose and the presence of will.

For a long time, biologists have been careful to exclude the "metaphysical heresy" of teleology from biological science. Most scientists and philosophers, including some Christians, have been skeptical about the idea of purpose in evolution, and some have claimed that any form of teleology is contrary to the very fabric of Darwinism. I think we need to stop "defending" the science of biology from the so-called danger of metaphysical or even spiritual encroachment. Let's welcome all the ideas of teleology, agency, design, and universal cognition into the fabric of our scientific endeavors, because we need these new ideas to help us move toward a new understanding of the essence of life.

The facts as demonstrated in chapter 4—that all of life, from its beginning, is full of teleology and willful agency, that cognition is present in the simplest of organisms, allowing for single cells to act with purpose—are now becoming visible and clear. These facts need an explanation, as does the well-known unique property of all life (chapter 3) to replicate itself accurately enough to allow evolution to begin to shape the rest of life, including something like me, who is writing these words, and you, who on reading them is thinking about the awesome nature of everything alive, including those you love.

And having accepted the truth of the previous paragraph, we should be able to see, in this new light, the outline of the missing key. We simply need to ask the crucial question, "Where did this purpose come from?" I believe there are two answers, which might actually both be part of the same answer. One is that there is something special about biology, a further hidden truth behind the existence of purpose, and with lots more exploration, that unknown principle, law, theory, or idea will come to light. The other answer is that we must either go outside of our science or expand (once again) the boundaries of what we call science to find the answer.

I strongly believe that both answers will lead to the same conclusion. The purpose that pervades all life will be traced to the creator of the universe. We won't find it by groping around in the darkness, but by seeing the missing key in the clear light of the beautiful, indeed holy, knowledge of biological science. Or saying the same thing in theological terms: the light of Christ.

Biology can also help us see that the idea *we don't need God to explain anything, so he needn't exist* isn't very scientific. Life contains many biosynthetic pathways complete with enzymes, control mechanisms, and energy requirements that at first glance seem redundant and unnecessary. It's as if some mother had said, "Take this sweater, just in case—you might need it!" And, sure enough, a sudden storm or cold front might prove the mother's wisdom. Our bodies might face a time when those "extra enzymatic pathways" become necessary. There are many things that, at first glance, we don't "need" in order to exist. But those things—love, beauty, and joy, to name just three—certainly exist. And so does God.

Of course, I do need God, and so do you. We need God to explain why we believe in God. We also need God to understand why we are here, and why there is love in the world. Why there is beauty, and why there is music. I sort of know how and why birds sing and why people can beat drums, play pipes, and sing tunes. But I need God to understand what happens to me when I hear Beethoven's *Symphony No. 9* or Franck's *Symphony in D Minor*, or anything by Bach or the Beatles.

I'll close by saying that I have no idea how long it will take until a brilliant young scientist will publish the first paper showing that one of the great mysteries of life, perhaps its origin, can be approached by a new law of biology that expands our knowledge of reality, much as all the established laws of chemistry and physics have done. I don't know how they will come up with this law, but as a matter of faith in the truth of science and that of Christ, I strongly believe that it will be the most magnificent pointer to the divine hand of our creator God that has ever been proposed. And then we will know that all of life declares the glory of God and fall to our knees in worship and praise.

And what a beginning that will be. Amen.

# Acknowledgments

I WANT TO THANK THOSE WHO HELPED get this book into your hands: my wife, Anikó Albert, who is my most reliable critic and editor; Dan Balow of the Steve Laube Agency; the wonderful team at Tyndale: Jan Long Harris, Jon Farrar, Jonathan L. Schindler, Kaylee Small, and the publicity team. Special thanks to manuscript beta readers Mark Meredith and Michelle Fauber.

The encouraging friendship of so many colleagues in the cause of science and Christian faith reconciliation has been an ongoing inspiration. Thanks to Denis Alexander, Joel Anderson, Zachary Ardern, Luke Barnes, Cameron and Brittany Bertuzzi, Justin Brierley, Francis Collins, Gregg Davidson, Tony Futerman, Oscar Gonzalez, Seth Hart, Eric Hernandez, Braxton Hunter, Michael Jones, Alister McGrath, Fuz Rana, Santiago Rangel, Paul Rimmer, Hugh Ross, Lee Strobel, Jim Stump, James Tour, John Van Sloten, Andy Walsh, Chris Wild, and Karen Wong.

I have had the honor of working with many wonderful Christians to spread the message of Christ as truth. I am grateful for the support given me by Dave Abernathy, Cheryl Bostrom, Eddie Bowling, Mike Clifford, Rebekah Davis, Ted Davis, Sheila Deeth, Kathy Emmons,

John Hall, Jana Harmon, Carmen LaBerge, Ashley Lande, Wilson Magaya, Perry Marshall, Sean McDowell, Jessica Moerman, Steve Moshier, Walter Roggero, Mario A. Russo, Sarah Salviander, Rick Saunders, Buddy Spaulding, Rebekah Valerius, and Patrick Woodward.

I am blessed to be a part of an organization devoted to science and the teachings of our Savior. The American Scientific Affiliation (ASA) is the largest organization in the Americas of Christians working in the sciences. The support of so many in the ASA has been an ongoing source of godly delight. I especially wish to thank my friends Kathryn Applegate, Paul Arveson, Mike Beidler, Vicki Best, Fred Cannon, Steve Contakes, Janel Curry, Becky English, Brian Greuel, Randy Isaac, Se Kim, Beth Madison, James Peterson, Preveen Sethupathy, Arnold Sikkema, Mark Strand, Josh Swamidass, Jennifer Wiseman, and John Wood.

My wonderful church, Rockville United Methodist Church, where I was baptized just over a decade ago and served as lay leader, has given me the blessings of true Christian fellowship and unwavering support. Thanks to Rev. Martha Meredith, Rev. Jane Wood, Rev. Vince Liburd, and all my friends in the congregation. Nothing is more important than family, so saving the best for last—thanks to Anikó, Edna, Cora and Yuri, Chaya, Giancarlo, Kevin and Ginika, Alex, and Emanuela for your love and support.

*Appendix*

# Technical Details

IN THIS APPENDIX, YOU WILL FIND MORE DETAILS about some of the material covered in the rest of the book. The purpose is not to provide a graduate level course in biochemistry but to satisfy the curiosity of those readers who want to dig a bit deeper into the wondrous world of the life of a cell and how it all works. Parts 1 and 2 present a fairly detailed and yet not-too-technical description of two of the most important and most complex biological machines in life.

The first machine is called the *ribosome*, a large set of cellular organelles whose function is the production of proteins. The second machine, called *ATP synthase*, is part of the process of converting energy from the sun or from food into chemical energy that can be used by the cell to power life. The much shorter part 3 is an extension of chapter 3 that covers more detail on my paper about self-replication as the fundamental process of biology.

## Part 1: The Translation Machine

*Translation* is the technical term for the conversion of the information in a cell's DNA into the proteins that control all the chemistry and structures of living cells. This process is also called protein synthesis since it describes how proteins are made in a living organism.

The translation process is quite a remarkable biochemical mechanism that takes place in ribosomes. It involves hundreds of steps and individual chemical reactions, as well as the work of some other machines. Most significantly, it operates using an informational instruction manual.

Ribosomes are complex structures made of RNA and proteins. They have two parts—a top and a bottom—and when these two parts come together to form an oval-shaped ball, a groove is formed between them.

In the days before personal computers and modern software, punch cards and sometimes long tapes of coded instructions were used to guide automated machinery. Some early computers worked this way. One can think of these methods as a primitive, non-electronic version of modern QR codes. The cards or the tapes were fed into the machine in a designated slot. The groove in the ribosome is the biochemical version of that slot, which is available for an instructional tape to be inserted.

### *The Instructional Tape*

After reading that last sentence, you might be thinking *Wait, what? Are we still talking about chemistry and molecules and living cells? What do you mean by "instructional tape"? Since when does a mere chemical give "instructions" to another chemical?* Great question, and the answer is coming. But remember: yes, this is chemistry, but it is biological chemistry, and the answer will take you far away from the known laws and rules of the rest of chemistry.

The "tape" that will go into the groove in the ribosome contains instructions, but it is not the ultimate source of that information, just as the computer tape is not itself the source of the instructions fed to the machine. In the computer's case, the information originates in a human mind—what the human wants the machine to do is written down, translated into code, and copied onto the tape.

In the biological case, the tape is also a copy of information that resides elsewhere. The source of the biochemical information is in the cell nucleus, a membrane-enclosed space that is home to the most remarkable molecule in the universe: the extremely long polymer we know as DNA.

### *DNA: The Information*

At first sight, DNA looks like a boring chemical. It contains two strands of units of phosphate and deoxyribose (a kind of sugar) molecules linked together, millions of units long. Each of the sugar molecules are also linked to one of four different chemicals we call nucleotides or bases, which appear to be arranged in a random order. Back in the early twentieth century, nobody thought that this seemingly nonfunctional polymer could have much significance, and certainly not that it could be the basic chemical of inheritance. They were wrong.

Scientists were already aware of the idea of a "gene" but had not yet been able to identify its chemical form. In 1953, James Watson and Francis Crick[1] discovered that if you look at the bases on the two strands of the DNA double helix, you always find an A (adenine) opposite a T (thymine), while a C (cytosine) is always opposite a G (guanine). They called this *base pairing*, and it is now known as a fundamental rule in molecular biology: A always and only pairs with T, and C always and only pairs with G.

The informational content of an organism's genes, stored in its DNA (or genome) is called the *genotype*. But in order for a cell to be alive, that information must be transcribed and translated into all the components of the cell, including all the proteins, lipids, and machinery that make the cell able to function. These components, and the kind of work (or functions) they do, are called the *phenotype*.

The existence of a genotype/phenotype linkage is a requirement for life. In all modern life that linkage—the engine that drives all living things—is produced by the protein synthesis translation system, the most complex set of biochemical processes that exist in biology.

The instructional tape is made out of RNA, whose sequence is copied from a gene made of DNA. A portion of the DNA double strand is separated, and the bases of one of the two strands are copied onto a single strand of RNA. The copying follows the base-pairing rule described above. Wherever there is a C in the DNA, the RNA will contain a G. Where there is a G in the DNA, we find a C in the RNA. For every T, there is an A, and wherever A appears, there will be a U (uracil, which is the RNA version of T). Base pairing is not a strong covalent chemical bond but a weaker kind of bond (called a hydrogen bond) that is easily formed and easily broken.

The relationship between each base and the one it forms a base pair with is called its complement. The messenger RNA strand is *complementary* to the DNA strand it was copied from, and since the two DNA strands are also complements of each other, the RNA strand is an actual copy of the other DNA strand.

Shown here are two hypothetical strands of DNA showing the strict base-pairing rule:

ATGTGGTTCACCAGGGTCAGTACT

TACACCAAGTGGTCCCAGTCATGA

And here is the sequence of RNA produced from the lower DNA strand above. Notice that its sequence is a perfect copy of the upper strand (only with U replacing T):

AUGUGGUUCACCAGGGUCAGUACU

The messenger (or mRNA) strand, which has all the information contained in the corresponding stretch of DNA, leaves the nucleus and carries the informational message of the DNA sequence to the ribosome, where it fits into the groove, just as a computer tape or punch card fits into the input slot.

Think of DNA as the book, genes as the chapters, the letters are the bases, and the words (made up of three bases/letters) are called *codons*.

And here is where the code begins. Each codon consists of three nucleotide bases in a specific order and has a meaning related to how to make a protein. The sequence of bases in a gene made of DNA means "this is the protein you should make—use the following sequence of different amino acids to make this particular protein."

### *Protein Structure*

The amino acid sequences of proteins are critical to allow every protein to carry out its function, whether as a catalyst for one of the thousands of chemical reactions needed for life to persist or as a structural element allowing for cellular integrity, cell divisions, and other functions.

Proteins are made of amino acids strung together in a long chain. The critical part of all proteins is the sequence of the amino acids, just as the critical part of all genes is the sequence of the bases. There are twenty different kinds of amino acids, and most proteins are about one hundred to three hundred amino acids long, meaning that the number of possible protein sequences is $20^{100}$ at a minimum, which is an inconceivably large number.

Now, you might think that if the bases in DNA form a sequence, and the amino acids in proteins form a sequence, it should be easy for one to directly produce the other. But life is never easy!

While the DNA bases and amino acids (as well as nucleic acids and proteins) can interact with each other, the specificity of these interactions (which amino acid binds to which nucleotide or base) is so low that it would be impossible to translate a DNA sequence of bases into one particular sequence of amino acids to form a specific protein.[2] As described in detail below, in the actual process, no amino acid interacts directly with the DNA or the mRNA at all.[3]

Because the DNA strand cannot directly organize the correct structure of a protein, the system needs to work with a set of chemical go-betweens or adapters. These include a translator protein enzyme, a kind of RNA called transfer RNA (tRNA), and many other molecules, which allow for the translation from one kind of chemical system (nucleic acids) to another (amino acids). It is a remarkable system, and the complexity of its workings far surpass any human-designed machinery. We can only marvel and wonder at its origin and profound ingenuity. But before we get into those details, let's go back to the code and unpack what it is and how it works.

### *The Genetic Code*

Nucleic acids and proteins are both polymers—long straight chains with no branches or forks. Both kinds of polymers can be made with no restrictions on sequence—in other words, there are no rules on which nucleotide or amino acid can go next to any other—and all possible orders of the subunits are equally possible.

There are only four different nucleotides, while there are twenty different amino acids. This means that a minimum of three nucleotides in a row are needed to code for a specific amino acid. (Two could only code for sixteen.) And in fact, that's what we have. Each group of three nucleotides (called codons) codes for one and only one amino acid. The identities of which codons code for which amino acid is what makes up the genetic code, which is generally presented in the form of a table (see below). In addition to coding for amino acids, there are three codons that stand for "STOP"—they tell the ribosome that the protein is finished and no more amino acids should be added.

To see how the genetic code works, we can use the same DNA sequence from above:

ATG TGG TTC ACC AGG GTC AGT ACT

Again, base pairing is not the genetic code—it is simply the chemical method by which DNA passes its information on to mRNA and also to the next generation of DNA.

I have divided the sequence above into groups of three bases; each group is a codon. In the table below, each of the possible three letter combinations of the four bases (U, C, A, G) codes for a particular amino acid (shown by a three-letter abbreviation on the right side of each box). As you can see, most amino acids can be coded for by several different codons.

Genetic Code Table

| First Letter | Second Letter: U | | C | | A | | G | | Third Letter |
|---|---|---|---|---|---|---|---|---|---|
| U | UUU | Phe | UCU | Ser | UAU | Tyr | UGU | Cys | U |
| | UUC | | UCC | | UAC | | UGC | | |
| | UUA | Leu | UCA | | UAA | Stop | UGA | Stop | |
| | UUG | | UCG | | UAG | Stop | UGG | Trp | |
| C | CUU | Leu | CCU | Pro | CAU | His | CGU | Arg | C |
| | CUC | | CCC | | CAC | | CGC | | |
| | CUA | | CCA | | CAA | Gln | CGA | | |
| | CUG | | CCG | | CAG | | CGG | | |
| A | AUU | Ile | ACU | Thr | AAU | Asn | AGU | Ser | A |
| | AUC | | ACC | | AAC | | AGC | | |
| | AUA | | ACA | | AAA | Lys | AGA | Arg | |
| | AUG | Met | ACG | | AAG | | AGG | | |
| G | GUU | Val | GCU | Ala | GAU | Asp | GGU | Gly | G |
| | GUC | | GCC | | GAC | | GGC | | |
| | GUA | | GCA | | GAA | Glu | GGA | | |
| | GUG | | GCG | | GAG | | GGG | | |

Codons cannot specifically bind to or interact directly with one and only one amino acid, at least not sufficiently strongly to avoid a high degree of error, which would make the system useless. So the physical manner in which the information on the mRNA strand (or tape) is translated into the correct amino acid sequence of a new protein is far more complicated than simply matching up each codon with the correct amino acid.

### *The Genetic Code in Depth*

To see what the genetic code is like, let's take a simple example. The DNA sequence AGGCTCACG is nine bases long, which means it contains three codons (there are no spaces in DNAese). The first word, AGG, means the amino acid arginine; the second word, CTC, means leucine; and the third, ACG, means threonine. These are three of the twenty amino acids, and all of them have at least one, and usually more than one, codon.

I should emphasize an important point here because people often get this confused: *the genetic code is not the DNA sequence*. The genetic code is the dictionary for the translation of the DNA sequence of bases in a gene into the sequence of amino acids in a protein.

I admit I find it annoying when someone says, "I love coffee—it's in my genetic code." No, it isn't, because we all (and by that, I mean all of life) have the same code, and I, for example, do not like coffee. There might be a *gene* that lets you and many others tolerate caffeine well. I would have another version of that gene, with a different *sequence*, which (when translated into a slightly different protein) makes me just a bit crazy with too much caffeine. The difference between us is due to our different *sequences* of the bases in that gene. Different DNA sequences are like different sentences written in the same language.

So half the requirement for a code is met—a symbolic meaning of the sequence of the bases in DNA contain information, similar to the language of a cookbook. But we need someone to read or "decode" that information and put it to use. So who is the recipient of that information? Who reads it, and what do they do with it?

The answer is the ribosome. A protein built in the ribosome is the final product of the information fed into the machine, just as the perfect chocolate cake is the result of the cook carefully following the written recipe. Now we are ready to peer into this machine and see how it works. We'll start by looking more closely at that mysterious tape carrying the information.

### *Anticodons*

Translation of the information in the DNA and mRNA into a protein involves much more than the mRNA and its codons. The essence of the system, based in ribosomes, includes base pairing as the "reading" mechanism. To understand this, I need to introduce one more term—the *anticodon.* As we saw above, every sequence of DNA has a complementary sequence based on the base-pairing rules. A three-base sequence that is complementary to a particular codon is called its anticodon. If the codon is ACG, the anticodon is UGC (in RNA). And we know that since the codon and anticodon are complements (by definition), the two sequences can easily form base pairs joining them together. It is the base pairing between codons and anticodons that form the mRNA "reading" mechanism within the ribosome.

### *RNA: The Working Nucleic Acid*

The NA in DNA stands for "nucleic acid," and as we saw earlier, there is another kind of nucleic acid called RNA. The differences between DNA and RNA are chemically minor: they have the same bases except for T, which in RNA is replaced by a similar chemical U (uracil). Also, RNA can assume many different kinds of shapes. There are actually three major kinds of RNA, and all of them are functional, whereas DNA only stores information.

As we've seen, messenger RNA (or mRNA) is a long straight molecule which serves as the instructional tape bringing the information from the DNA to the ribosome. This implies that there has to be a mechanism that copies the information (the sequence of bases) from the DNA to the RNA. In other words, the information must be *transcribed*, analogous to someone typing up the words of a speech to be read by others. Transcription is, in fact, the technical name for the process of copying the genetic information in the base sequence from DNA to RNA.

### *Transfer RNA*

In the ribosome there are two open regions that allow other molecules to enter the ribosome and make contact with the mRNA in the groove. These open pockets are called the A site and the P site.

The molecules that fit into the ribosome pockets at these two sites are another kind of RNA called transfer RNA, or tRNA. These are a family of molecules made of the same nucleotide components as mRNA, but they are quite different in their structures. What makes tRNA so different from mRNA? Their shapes. While mRNA is a long chain, the sequences of all tRNAs are arranged in three dimensions in such a way that they fold into various shapes. The way they do this is by base pairing within the tRNA sequence so sections that are complementary bind together forming loops, folds, and the creation of specific shapes reminiscent of proteins.

The shapes of the different tRNAs are important in what comes next. The detailed structure of all tRNAs have two specific parts of the molecule in common. At the top of each is a short single strand, which binds a specific amino acid. At the bottom of the folded tRNA molecule there is a three-base section called the anticodon region. It is this section of the tRNA that interacts with the codons on the mRNA.

### *About the Ribosome Machine*

We can now get back to the ribosome machine, where the tape of mRNA carrying the information from the gene is about to enter that groove in the middle of the machine. The mRNA fits into the groove and begins to move through the ribosome exactly one codon at a time. The ribosome is now "reading" the information on the mRNA tape.

When I say the ribosome reads the tape, what do I mean? Let's go back to our human-built tape-reading machine. The holes in the punch cards or the images on the tape exert an effect on the mechanisms of the machine, much as a phonograph produces music by the mechanical translation of the record's imprint into specific sounds or the way striking the keys on a typewriter produces the appearance of a letter on a paper. (I know I am dating myself here.) The same kind of thing happens when the mRNA tape moves through the ribosome one codon at a time.

Every specific kind of tRNA molecule has bound to it one and only one specific kind of amino acid. What we need to know now is that it's the tRNA, not the amino acid itself, that interacts with the base sequence on the mRNA tape. Remember that amino acids and RNA bases do not interact well with each other, so the tRNA acts as a carrier of the amino acids to go to the right place on the mRNA tape.

### *The Ribosome Machine at Work*

Now let's see how the tape is read in the ribosome. For every codon on the mRNA tape, a corresponding molecule of tRNA attaches to it. This is analogous to a human-built machine with a conveyor belt to which workers add specific items that match what is already on the belt. Of course, in the ribosome there are no workers, and the selection of what items (tRNAs) to add is done automatically by chemical interactions between the codon on the mRNA tape and the right tRNA.

Again, we will use the example given above of a small section of DNA. But now, writing the codon sequence in RNA, we have

AUG UGG UUC ACC AGG GUC AGU ACU

When translation starts, the mRNA enters the groove between the two parts of the ribosome, and the first codon on the mRNA is at the A site. The mRNA then moves, one codon at a time, through the ribosome. When the mRNA tape moves, the first codon passes from the A site into the P site and the second codon is at the A site. At the next move the first codon is now outside the ribosome, the second codon is at the P site, and the third codon is at the A site. And this continues as the mRNA moves through the ribosome.

Of course, a lot more happens than just the mRNA moving through the ribosome. The information on the mRNA tape must be read and translated. It starts with the first codon on the mRNA at the A site. The tRNAs can enter the A site and form a bond with the codon on the mRNA. This is done by base pairing between the codon (on mRNA) and the anticodon on the tRNA. So if the codon is AUG, then the tRNA with the matching complementary anticodon (in this case UAC) will enter the A site and base-pair (the A with the U, the U with the A, and the G with the C) with the mRNA. No other tRNAs can bind to the mRNA at the A site at that time—only the one with a UAC anticodon.

The tRNA that has the right three-base anticodon sequence to bind to the codon of the mRNA at the A site also happens to have just the right amino acid attached to it. In this case that amino acid is methionine. In other words, only a tRNA with methionine on it can bind to an mRNA with the codon AUG, which is why that codon is the code for methionine.

The same is true for every three-base codon sequence. Each tRNA binds specifically to one codon in mRNA. And every tRNA with an anticodon that matches any of the codons for a specific amino acid carries this amino acid with it. *How does that happen?* you might ask. You will find out shortly.

All mRNAs start with the codon for the same amino acid—methionine. (In most proteins this is removed later.) The ATG codon in DNA is the "start" signal for making a protein, and AUG is the codon for methionine and is the first codon of the mRNA sequence. All tRNAs carrying the amino acid methionine at the top have the sequence UAC in their anticodon region.

Once the first tRNA with the anticodon sequence UAC base-pairs with the first codon (AUG) on the mRNA tape with a methionine in the A site, the tape moves three bases over, exposing the next codon: UGG. That is the codon for tryptophan. Only a tRNA that has the sequence ACC at the anticodon site (the complement of UGG) can bind to that second codon on the mRNA. And every tRNA with ACC at the anticodon site has the amino acid tryptophan sitting on top of it.

When the tryptophan-bound tRNA base-pairs with the mRNA codon for tryptophan, we have two amino acids, methionine (at the P site) and tryptophan (at the A site), sitting on their respective tRNAs in the ribosome machine. And now the third kind of RNA, ribosomal RNA (rRNA), in that machine gets into the act by catalyzing a chemical reaction that breaks the bond between the first amino acid and its tRNA and binds the two amino acids together in a strong covalent bond called a peptide bond. This is formed by a bond between the carboxylic acid group at one end of the first amino acid and the amino group at the other end of the second amino acid. The ribosomal RNA that catalyzes this chemical reaction between each amino acid and its neighbor in the ribosome is called a *ribozyme* (discussed in Chapter 5).

When this reaction is consummated, the first amino acid (methionine) is no longer bound to its tRNA, and that tRNA leaves the ribosome. The mRNA tape again moves three bases, so the second amino acid, tryptophan, goes from the A site to the P site, and the third amino acid, phenylalanine, moves with its tRNA into the A site. The ribozyme then attaches the third amino acid to the second, and we now have a three amino acid chain, the beginning of a protein.

The process repeats as the mRNA tape moves through the ribosome, adding another amino acid to the growing chain until all 100 to 400 codons have moved through, and all the right amino acids corresponding to those codons have joined the chain. When the last codon, one of the three STOP codons gets read, the protein is released into the cell to do its intended job as an enzyme or whatever else it was made for. And that is how proteins with the just-right sequence—coded for by the sequence of bases in the DNA, transcribed into the sequence of the mRNA, and translated by the machinery of the ribosome using the tRNA tools—are made.

Figure A.1 shows the process in pictorial form, which might help you to grasp the amazing beauty of this basic machinery of life.

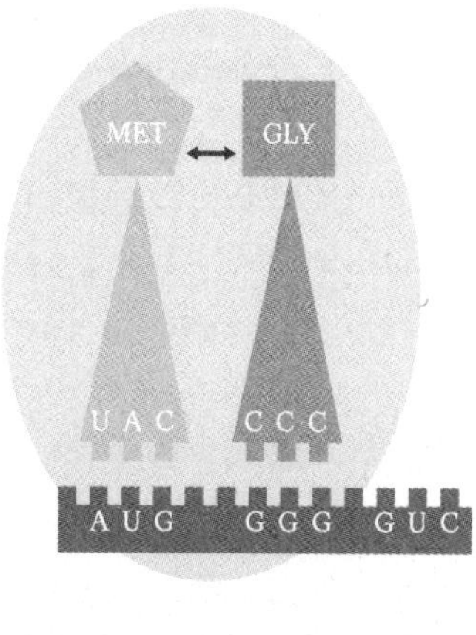


**FIGURE A.1** *The first two amino acids bound to their tRNAs go to the first two positions on the mRNA in the ribosome, because of base pairing between the codon on the mRNA and the anticodon on the tRNA. A covalent bond between the two amino acids is catalyzed by the ribosome.*

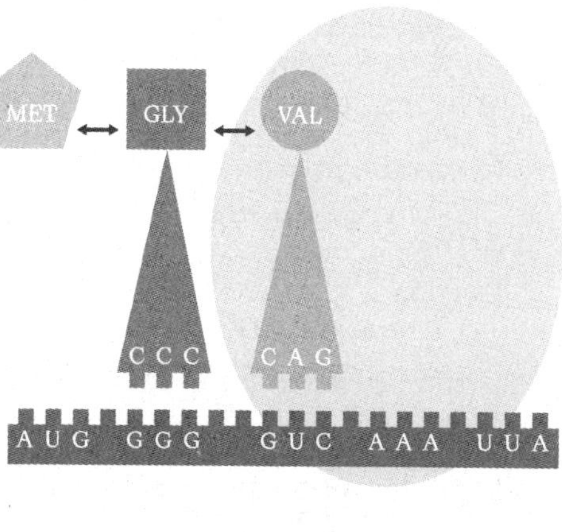


**FIGURE A.2** *The mRNA tape moves along the ribosome, adding tRNAs and amino acids to a growing protein chain.*

### *The Translator*

Now it's time to answer the critical question: How does the tRNA with just the right anticodon know which amino acid to bind to? After all, it cannot just look up the answer in the genetic code table. And there is no direct chemical reason that a particular amino acid bound on one end of the tRNA molecule would signify that the other end should contain the correct anticodon sequence to bind to the correct codon on mRNA. So, how does that happen?

The answer to that question, if it were asked about a manufacturing process, would probably be that there is another process where well-trained technicians attach each amino acid to the right tRNA, and perhaps there is an inspector to make sure that no mistakes were made. And that's what happens in cells—except that, of course, there are no tiny people doing the work. Instead, there is a remarkable enzyme (or set of enzymes), called *aminoacyl-tRNA synthetase* (*aaRS*).

The most basic and smallest of all the molecular machines in living cells are the protein enzymes themselves. Some are relatively simple, and some are highly complex, and aaRS is one of the latter. This enzyme is what puts every amino acid onto the correct tRNA in a process that is almost as complex as the ribosomal machinery we just saw.

Aminoacyl-tRNA synthetase is at the very center of the translation process, and might even be considered the actual translator, the chemical entity that speaks two languages—nucleotidese and amino acidese—and can translate each into the other. It is this enzyme that links up two chemicals that otherwise would have no chance of getting together: an amino acid and a transfer RNA.

The aaRS is not one enzyme but a family of related enzymes, with different molecules specific for each of the twenty amino acids. The name *aminoacyl-tRNA synthetase* is the name of this family, but each individual member is named for the amino acid it is responsible for, such as arginine tRNA synthetase, proline tRNA synthetase, and so on. These enzymes do not read the genetic code—they are the entities that *make* the genetic code what it is. They do that by binding both one specific amino acid and one kind of tRNA with a particular sequence in its anticodon region. This means that the only way to change the genetic code would be to have mutations that change the specific binding of amino acids and tRNAs to these enzymes, and if you want to go more deeply into the biochemistry of that, you will see that it's next to impossible. This is why the genetic code has not changed (except for extremely minor variations) since the dawn of life and is universal since the last universal common ancestor (LUCA).

Figures A.3–A.5 gives a rough idea of how this enzyme carries out its function.

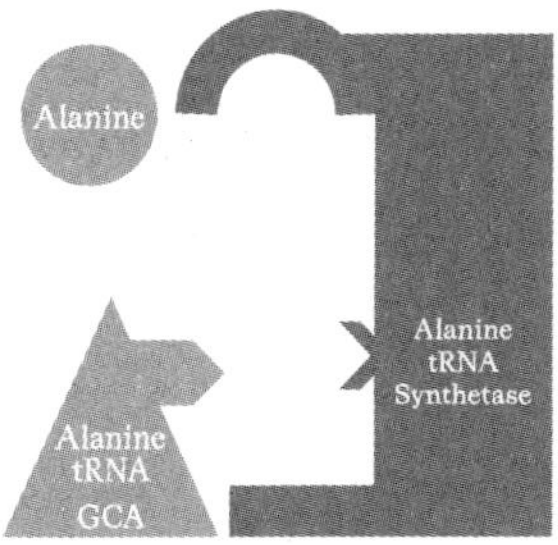


**FIGURE A.3** *Aminoacyl-tRNA synthetase, the amino acid alanine, and the tRNA specific for alanine, with an alanine anticodon.*

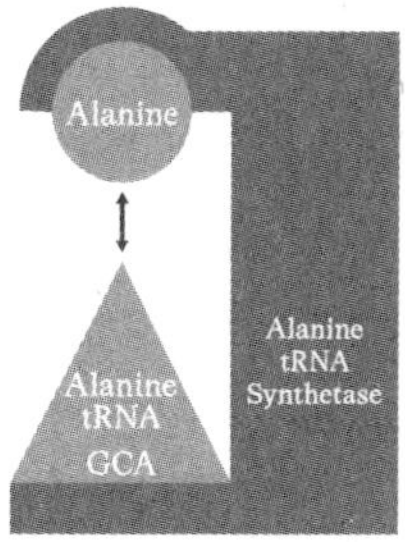


**FIGURE A.4** *Alanine and its tRNA bind to the enzyme and are joined by a chemical reaction.*

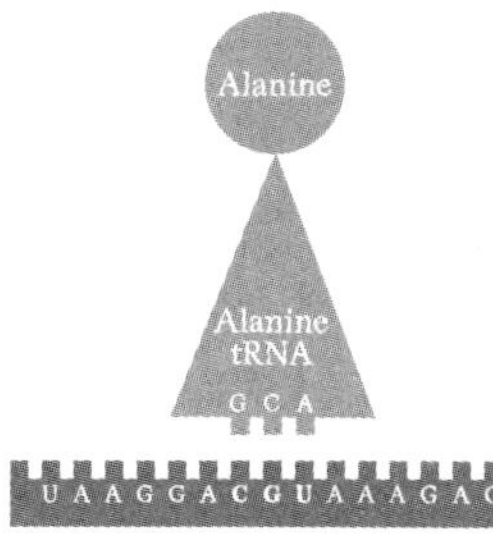


**FIGURE A.5** *The alanine bound to its tRNA leave the enzyme and travel to the ribosome, where the codon-anticodon binding places the alanine in the correct position.*

Let's take a closer look at how the translator molecule functions. Each enzyme has a binding site for its specific amino acid. So alanine RS has an alanine binding site. It also has a binding site for each of the four tRNAs that carry any of the four anticodons for alanine. Once this very large enzyme has an alanine bound in one site and an alanine tRNA bound on the other site, the protein conformation (its shape) shifts in a way so that the end of the tRNA and one end of the amino acid are close. The enzyme then catalyzes a chemical reaction (requiring energy) between the two molecules, and the result is a strong (covalent) chemical bond between the tRNA and the amino acid.

The enzyme has done its job. It has taken its amino acid and matched it with the right tRNA so that the amino acid will be put in exactly the right place determined by the base pairing of the anticodon site on the tRNA with the codon on the mRNA. Once this is done, the new bonded tRNA amino acid complex can leave the enzyme and as a tightly bound couple navigate its way to the ribosome.

But that is not the whole story of aaRS. The entire process of translation needs to be very accurate. If the wrong amino acid gets attached to the wrong tRNA, then the protein sequence will not be what it's supposed to be and the protein might not work. In fact, the entire translation process, including the work of the aaRS, is extremely accurate, with only one mistake in over ten thousand trials. And to reach that level of accuracy the binding and matching I described above just won't cut it.

One problem is that many amino acids are very similar in structure to many other amino acids and mistakes can easily be made. A leucine might just fit into the binding site on an isoleucine or a valine aaRS, and that wouldn't be good. And alanine is only slightly bigger than glycine, and it would take an amazingly well-engineered binding site to be able to distinguish the right from the wrong amino acid with 99.999 percent success.

So the enzyme has some tricks that allow it to function as the inspector checking for errors as well as the technician putting the two items together. Once the amino acid and tRNA are bound together by the enzyme, the happy couple is shunted to another site on the same enzyme called the editing site. This site is shaped in a way that will allow almost any amino acid and tRNA pair to get into it, except the right one. The correct amino acid/tRNA combination just doesn't fit.

It's especially good at admitting smaller amino acids, while excluding the right one. So if everything is correct, the amino acid/tRNA complex does not—*cannot*—bind in the editing binding site, and it breaks free from the enzyme to go on its merry way toward the ribosome. But if a mismatch or some other error has taken place, and the mistakenly bound couple *does* fit into the editing site, it's cut into pieces by the action of the enzyme and never gets close to the ribosome.

So not only does this enzyme recognize the proper amino acids and from one to six proper tRNAs, and not only does it provide the energy and mechanical means to link the two together, but it also makes sure it got it right and deals a death blow to any erroneous products.

Now take a deep breath while I tell you that every cell in every living thing on this planet has the same system—and *has* had it as far back as LUCA.

As you might imagine, we have no idea how this could have originated. No human mind could have conceived of this design and put it into action. And as we saw in chapter 3, it could not have possibly evolved through the evolutionary process we know and understand, because that process can only get started once this translation machinery is working with high accuracy and efficiency.

There is good evidence that the code we know today evolved from more primitive forms, which might have had codons made of two bases instead of three. This early form of the code would only be able to allow for proteins composed of sixteen (or fewer) amino acids, as opposed to the twenty used in modern life. But that does nothing to counter the argument that the origin of any kind of code—even a simpler one—could not have originated through evolution.

I made some simple videos that might be helpful companions to this appendix.[4] You can also find many excellent, more professional and detailed videos online that show graphic illustrations of the translation from genetic code to protein synthesis by the ribosome. Most of them show only parts of the process, but it pays to watch a few to illustrate the detail I have presented here. And for even more detail, advanced textbooks like *Molecular Biology of the Cell*[5] or *Lewin's Genes XII*[6] are highly recommended.

## Part 2: Other Biochemical Machinery

I think the protein synthesis machinery is the most impressive feature in all of life, and probably in the whole universe, with the possible exception of human consciousness. But it is far from the only machine of this kind in every cell of your body. The others include several machines in the mitochondria that are involved in energy conversion; the receptors and portals in cell membranes that allow certain molecules to pass through in one direction but not the other, or allow other molecules to bind to proteins causing a series of chain reactions that can lead to major changes in how the cell behaves; the mechanical cables and switches that allow the cell and the nucleus to divide in two; and many others.

I will give a glimpse into one of my other favorites, a machine made of many proteins called *ATP synthase*. It's in charge of the final step in a long and complex process of turning the energy acquired from eating food (or from sunlight in plants) into the chemical energy that powers almost all the chemical reactions in the cell that require energy input.

### *Energy*

Energy is everywhere in the universe, often obvious, as in the burning of stars, or less so, as in the gravitational energy between distant bodies or the chemical energies in the atmospheres or on the surfaces of planets. For anything that happens, anywhere, energy is either released or required.

Life needs energy to exist and function. You can cram a cell with all the stuff you want. Proteins, DNA, lipids, whatever. You can even grab some ready-made structures from living cells like membranes, tubules, all the complex molecular machines you can find, and stuff them inside and hook them up, but without energy, nothing happens—no life. It would be like a state-of-the-art modern factory with the power turned off.

When physicists say that matter and energy are the same thing and one can be turned into the other, that sounds very nice, and it's pretty much what you might expect a physicist to say. But down in the real world of worms, amoebas, and living cells, it just doesn't happen. Sure, you can turn matter into energy if you have a nuclear bomb handy or if you are hanging out inside a star at 10,000,000 degrees. But on earth, nope. Matter turning into energy is way too hot for any kind of life.

So where does life on earth get the energy it needs? From the sun. All the energy that powers all of life is solar. We can think (and many have) about the fortunate coincidence that stars are tremendous generators of both heat and light energy, while at the same time strong sources of gravitational energy, thus drawing planets close enough to them to be showered by their energy fields. We can think of stars as the nurturers of potential life. The opposite is true for the greatest sources of gravitation—black holes, which emit no useful energy and engulf all objects that are pulled too close.

There is a never-violated law of physics that says that energy cannot be created or destroyed but it can be converted from one form to another. That is very good news, because life requires many types of energy—chemical energy, mechanical energy, heat energy, electrical energy, hydraulic energy, osmotic energy, and lots of variations on these. And the fact that one kind of energy—the kind we get from the sun—can be converted into all these other kinds makes life possible.

The sun bathes the earth in a constant shower of energy that comes in two forms—heat and light. While it's nice that the sun's radiant heat keeps the earth and all who live on it warm and toasty, what provides the energy used by life isn't solar heat but sunlight. If you have solar energy panels on your home, those panels collect light energy and convert it to electricity. And that's what the green plants on our planet do also, except that the conversion process is far more complex and far more efficient.

But all of that, while wonderful and thrilling, is not the whole story. Neither we, nor our cats, dogs, or the squirrels in our yards or the birds in the sky or the fish in the sea have anything to do with converting sunlight into the energy we all need to live. Instead, what all animals do is eat the organisms that can do that. And then some of us eat the other animals that eat the plants. So our energy comes from food, and that food ultimately comes from the living creatures that made it using the energy from the sun.

The food we eat (which is matter) is not converted by any kind of thermonuclear reactor in our bodies, although that would make a good science fiction theme. That food is energy rich, and the chemical energy in food is converted by animals into all the other kinds of energy needed for survival. I keep talking about energy conversion, but what is that all about? How is it done, and how does it all work? The answer, as you might have guessed by now, is all about some biochemical machines.

### *The ATP-Making Machine*

As the name suggests, the purpose of the ATP synthase machine is to make the molecule ATP, which stands for adenosine triphosphate. Why is that important? What's so special about ATP? It's those three phosphates at the end of an adenosine molecule (the same adenosine that is found in DNA and we abbreviate as A in the genetic code). Phosphate is made of a phosphorus atom linked to four oxygen atoms, and triphosphate is made of three of those linked together by high-energy covalent bonds. This means it takes a lot of energy to make a triphosphate, and when one of those bonds breaks, a great deal of energy is released. That is why we find ATP involved with almost all cellular biochemical reactions that need energy. When one of the phosphate bonds is broken (usually by an enzyme as part of its catalytic action), producing adenosine diphosphate and a single phosphate, the energy released can power a biochemical reaction.

### *Prelude to Making ATP*

To begin, I need to say that there are many, many steps before we can even get to the ATP-making machine. Some of those steps involve a series of chemical reactions that go in a cycle and produce high-energy chemicals and hydrogen atoms (from water and food) as well as free electrons. Hydrogen is number one in the periodic table and consists of a single *proton* in the nucleus and a single *electron*. Both of those subatomic particles are crucial for making ATP. The steps needed before we get to actual ATP production are described in biochemistry graduate textbooks as the electron transport chain (ETC).

I will not discuss it at all here, because it is so complex that even if I tried an oversimplified version, you would throw the book across the room (I am assuming you haven't already done so).

The ETC does all the things needed for the ATP synthase machine to work. First, it uses those electrons generated from food and water to produce energy, and then that energy is used to drive all the protons (also generated by the ETC) across a membrane.

### *Potential Energy, Protons, and Membranes*

You might ask: *What? All those reactions, all those molecules and enzymes, and the end result is pushing a bunch of protons across a membrane? That's it? What is the big deal about having so many protons on one side of a membrane?*

As you will see, it's a very big deal, and the reason is potential energy. What does it mean when we say something has potential energy? It means the thing, or the system, isn't actually doing anything energetic, like when an axe is brought down to chop a piece of wood—but it can. And the reason it can is because energy has been put into it (like raising the axe) that is stored for now but can be released into its environment in the right circumstances. The axe lying on the ground has no potential energy, but when I lift it over my head (which requires energy), it does. That potential energy turns into kinetic energy when I bring the axe down onto the wood.

Life works that way when it's time to make ATP. All those tiny protons sitting on one side of a membrane are full of potential energy. Just like the axe poised over my head, ready to strike, the protons are a potential force. If a hole opened in the membrane, the protons would rush through it in one direction, since the laws of thermodynamics insist that any collection of molecules or particles or any other bits of matter or energy must be distributed evenly in a container. And there *is* a hole in this membrane, but it's a very special hole. It can be opened and closed, and it allows only a narrow, regulated stream of protons to go through. And most importantly, the hole is part of a large and astonishing machine: ATP synthase.

### *The Ultimate Biochemical Machine at Work*

On the other side of the hole is a complex set of proteins arranged in a cylinder. As each proton goes through the hole, it interacts with one of those proteins and gives it a push. Since we are talking about thousands of protons per second crossing the membrane through the hole, the result is that the cylinder turns. In fact, it turns at a rate of about 6,000 rpm, or one hundred revolutions each second. That is *six times* the maximum speed of a turning wood lathe.

This motor, called F0, is a work of enormous ingenuity and craftsmanship, but it's only the beginning. Motors have purposes, and the function of the F0 motor is to turn a drive shaft (also made of proteins), which is connected to another motor called F1.

There are still some mysteries about how F1 works, but we do know that the proteins in F1 are arranged in a complex, asymmetric way, so that as it turns at this high speed, there are points in its rotation when a molecule of ADP (adenosine linked to two phosphates) and a molecule of phosphate can bind at the same time very close together to one of the proteins. Then, as the rotation continues, the positions of the amino acids in these proteins shift in a way that pushes the single phosphate onto the second of the two phosphates on the adenosine, and a chemical bond is forced to form. The mechanical energy of the turning F1 motor is used to power the bond formation, resulting in a bond with high chemical energy, which can later be used to power the many kinds of cellular chemical reactions that require energy to go.

And thus, a new molecule of ATP, the universal chemical energy carrier in all of life, is born. There are many online videos of ATP synthase at work. Your homework is to watch at least one of these.

## Part 3: My Paper on Accurate Self-Replication

In chapter 3, I mentioned a paper I published in the journal *BioSystems* that demonstrates mathematically that high-accuracy self-replication is the fundamental feature of life that allows evolution to occur. Most of the focus of the paper is about deriving a simple equation that shows the existence of the threshold for replication accuracy. That equation is:

$$\Delta p = \frac{p_{n-1}(F-(\overline{w}-Y))}{\overline{w}-Y}$$

Since evolution is defined as a change in allele frequency over time in a population, the term $\Delta p$ ("delta p," which means the change in the frequency of allele "p") must be greater than zero for evolution to happen. If you look at the equation, you will see that if the term $(F-(\overline{w}-Y))$ is equal to 0, then the entire right side of the equation will equal 0, which means that $\Delta p$ would equal zero. So in order for evolution to happen, the term $F$, which is the degree of accuracy of self-replication, must be greater than the threshold $\bar{w} - Y$.

The term $\bar{w}$ stands for the average fitness of all the alleles, and $Y$ stands for the error rate times the initial frequency of the $p$ allele. So what this equation says is that the degree of accurate replication must be greater than the difference between the average fitness of all the alleles and the product of the error rates times the initial $p$ allele frequency.

# Glossary

THE DEFINITIONS OF MANY OF THE SCIENTIFIC TERMS used in this book are given here. The list is not exhaustive, and the definitions are basic and avoid overly technical language wherever possible. Further information about any of these terms can be found online or in many of the reference notes. Terms in italics are defined in the glossary.

**aaRS (aminoacyl-tRNA synthetase):** an *enzyme* that combines one kind of *transfer RNA* that contains an *anticodon* for a specific *amino acid* with that amino acid. There are different aaRS molecules for the different amino acids. They are the molecules responsible for the *translation* of the *genetic code* written in the language of *DNA* into the sequence of *proteins.*

**abiogenesis:** the origin of life from nonlife. The details of this process are unknown, and the assumption by many scientists that life began as a result of only the known laws of chemistry and physics has not been shown to be correct.

**agency:** taking action based on the will or volition of a living creature. Agency is a universal property of life and does not occur in the absence of life.

**allele:** most *genes* include small variations (produced by *mutations*) in the *DNA* sequence found in different individuals in the same species. A gene can have anywhere from one to dozens of different alleles, which may or may not have an effect on the function of the gene. For example, different responses to medications can originate from different alleles of the same genes.

**amino acid:** a molecule composed of carbon, oxygen, and nitrogen atoms that forms the basis for the long polymers called *proteins*. Each amino acid has an acidic end and a basic end, which allow them to form linkages with each other in long straight chains. There are twenty different kinds of amino acids in all of life.

**anticodon:** a sequence of three *nucleotides* (or *bases*) in *DNA* or *RNA* that are complementary to the three bases in a *codon*. For an explanation of "complementary," see *base pairing*.

**ATP:** adenosine triphosphate is the universal carrier of energy in all biological organisms. ATP is derived from the conversion of solar light energy into chemical energy by *photosynthesis* in plants and by the metabolism of energy containing molecules in food by animals.

**base:** bases are the central components of nucleic acids. They are made of one or two rings of carbon atoms with other atoms attached. The four bases in *DNA* are adenine (A), cytosine (C), guanine (G), and thymine (T). The first three are also present in *RNA*, but thymine is replaced by uracil (U) in RNA. The sequence of these bases in DNA is the source of the information in *genes* on how to build all the proteins in a cell.

**base pairing:** strands of *DNA* or *RNA* interact due to the weak chemical attraction of certain *bases* with each other. This attraction is produced by weak hydrogen bonds between specific bases. The two-ring base A always and only pairs with single-ring T (or U in RNA), and the two-ring base G always and only pairs with single-ring C. This specific base pairing is the key to understanding many aspects of molecular biology, including how *genes* work.

**catalysis:** a process in which a chemical agent assists in a chemical reaction without being affected by it. Generally, catalysts enable a much faster reaction time. In life, the vast majority of catalysts are *enzymes* made of *protein*. The action of catalytic enzymes allows biochemical reactions to reach completion thousands of times faster than they would without a catalyst.

**codon:** a sequence of three bases in *DNA* or *RNA* that codes for one specific *amino acid*. There are sixty-four possible codons of the four bases arranged in groups of three. One to six codons code for each amino acid. The identity of which codons code for which amino acids is the same throughout all of life (with a few very minor variations) and is called the *genetic code*.

**DNA:** deoxyribonucleic acid is a double-stranded polynucleotide polymer that makes up all the *genes* in a cell. Most of the DNA is outside of the small sections that are the actual genes. Some of this DNA has other functions such as regulating the expression of a gene or providing material for new genes. Some of the nongenic DNA is nonfunctional, although the percent that might be called "junk DNA" is controversial.

**enzyme:** a *protein* molecule that is highly specific as a catalyst for one specific chemical reaction. Enzymes act with extremely high speed and efficiency. They can also provide the energy to drive energy-requiring reactions.

**gene:** the basic unit of inheritance, whose structure and functional biochemistry was elucidated in the 1950s and '60s. Genes are composed of a small *DNA* sequence that codes for a specific *protein*. The detailed structures of genes are complex, usually consisting of coding regions (exons) and noncoding regions (introns). Genes may be expressed (meaning they actively code for the production of new proteins) or unexpressed, and the regulation of gene expression is a complex field of biochemistry.

**genetic code:** all of the sixty-four possible combinations of the four bases in groups of three (*codons*) have a functional significance. One codon acts as the start of a new *protein*, two signal the end of the protein chain, and each of the other sixty-one codons code for one *amino acid*. The genetic code is the sum of all the correspondences between the codons in *DNA* and the amino acids they code for. There is only one genetic code in all of life (with minor exceptions), and it has not evolved since the *last universal common ancestor (LUCA)*.

**genotype:** the totality of all the *genes* in an organism. This includes coding and noncoding regions, regulatory elements, and repetitive sections. All the members of a particular species share most of the same genotype (or genome), but individuals usually have small differences in some specific gene sequences called *alleles*.

***Homo sapiens:*** the species of modern human beings, a member of the genus *Homo*. Although fossils of extinct species of *Homo* are known, the most recent species—*Homo neanderthalensis*—went extinct about 35,000 years ago. Due to genetic bottlenecks and other factors, *Homo sapiens* is more genetically homogeneous than most other species in the primate family.

**last universal common ancestor** (LUCA): a hypothetical organism that existed before any branching of the tree of life (phylogenetic tree) took place. Its characteristics and properties are based on those *phenotypic* features that are present in all branches of modern or known previous life forms. Examples include a genetic code, the protein synthesis system, and a complex bilayer membrane with receptors and portals. LUCA is not the first life form (about which we know very little) but the last living creature that gave rise to all life as we know it.

**messenger RNA (mRNA):** see *RNA*.

**mutation:** a copying error in the replication of *DNA* caused by chance or the action of certain chemicals or radiation. The resulting error in the DNA sequence in a *gene* or regulatory region can lead to a nonfunctional *protein* or a protein with a different function, although the majority of single-*base* mutations have little if any effect on protein function. There are various types of mutation in addition to point mutations, where a single base is replaced by another base. These include frameshift mutations, large deletions, and the insertion of new DNA carried by viruses.

**nucleoside:** the monomeric units of *DNA* and *RNA* polymers. In addition to one of the four *bases*, each nucleoside contains a sugar—deoxyribose (for DNA) or ribose (for RNA). The nucleosides are called adenosine (A), cytidine (C), guanosine (G), and thymidine (T). In RNA, thymidine is replaced by uridine (U).

**nucleotide:** a nucleoside with a phosphate attached to the ribose or deoxyribose molecule. The polynucleotide chains (*DNA* and *RNA*) are produced by bonds between the phosphate of one nucleotide and the ribose of the next nucleotide in the chain. The nucleotides are named using the nucleoside name plus the number (from one to three) of phosphates attached. For example, the nucleotides of adenine are adenosine phosphate, adenosine diphosphate, and adenosine triphosphate.

**phenotype:** all characteristics of a cell or organism, including its structural and functional aspects, as well as all the biochemical properties and behaviors of the cell or multicellular organism.

**photosynthesis:** the process in green plants that converts solar light energy into sugar and chemical energy. All life depends on this process for the production of food and energy.

**protein:** a polymer composed of *amino acids* in a specific sequence that reflects the sequence of the *bases* in a *gene*. Proteins may be catalytic *enzymes* or structural elements such as membranes, filaments, and organelles. Most proteins function as parts of biochemical machines, and enzymes are themselves miniature machines that manufacture all the components of living cells.

**ribosome:** an organelle that is the site of protein synthesis. Ribosomes are composed of *RNA* and *protein*, which form a football-shaped structure with a groove between the top and bottom subunits. Messenger RNA, produced in the nucleus and carrying a copy of a *gene* sequence, enters the groove, where each *codon* forms a *base-pairing* bond with a transfer RNA carrying an *amino acid*. The ribosome then links adjacent amino acids together to form a protein chain consisting of amino acids in just the right order.

**RNA:** a polyribonucleic acid, like *DNA*, with two differences. The sugar component is ribose rather than deoxyribose, meaning that RNA has an extra oxygen atom. Also, RNA uses uracil instead of thymine as one of its *bases*. RNA comes in three structurally different forms. Messenger RNA (mRNA) has a complementary sequence (see *base pairing*) to the DNA sequence of a *gene* and is a long single strand. Transfer RNA (tRNA) is a smaller folded molecule with an *anticodon* at one end and a binding site for an *amino acid* at the other end. Ribosomal RNA (rRNA) is part of the structure of the *ribosome* and has *enzymatic*-like activity, able to *catalyze* the bonding reaction between two amino acids.

**teleology:** the "final cause" in Aristotelian philosophy, meaning the reason things happen or the purpose behind an event. Modern science has ruled out the existence of teleology in physics and chemistry, but the presence of purpose in the science of life is currently controversial.

**transfer RNA (tRNA):** see *RNA*.

**translation:** the mechanism of *protein* synthesis involving the complex machinery of the *ribosome*. It is perhaps the most complex biochemical system in biology and requires the actions of *messenger RNA*, *transfer RNA*, *amino acids*, *aminoacyl-tRNA synthetase (aaRS)*, and many other biochemicals. The system converts the information contained in the sequence of *bases* in *genes* into the sequence of amino acids in proteins.

# Notes

## INTRODUCTION: HOW DID WE GET HERE?

1. See Sy Garte, *The Works of His Hands: A Scientist's Journey from Atheism to Faith* (Kregel, 2019).
2. Ruth Bancewicz, "Can Science Prove God Exists?," The Faraday Institute for Science and Religion, February 20, 2020, https://www.faraday.cam.ac.uk/churches/church-resources/posts/can-science-prove-god-exists/.
3. Richard Dawkins, *River out of Eden: A Darwinian View of Life* (Basic Books, 1995), 133.
4. The revolutionary new physics and its implications for belief in God are beautifully described in Michael Guillen, *Believing Is Seeing: A Physicist Explains How Science Shattered His Atheism and Revealed the Necessity of Faith* (Tyndale Refresh, 2021).
5. Hendrik Antoon Lorentz, Albert Einstein, Hermann Minkowski, and Hermann Weyl, with Arnold Sommerfeld, *The Principle of Relativity: A Collection of Original Memoirs on the Special and General Theory of Relativity*, trans. W. Perrett and G. B. Jeffery (Dover, 1952).
6. Sy Garte, *Science and Faith in Harmony: Contemplations on a Distilled Doxology* (Kregel, 2024).
7. Justin Brierley, *The Surprising Rebirth of Belief in God: Why New Atheism Grew Old and Secular Thinkers Are Considering Christianity Again* (Tyndale Elevate, 2023), 164.

## CHAPTER 1: BIOLOGICAL IDEOLOGY AND A CALL FOR UNITY

1. Mary Domski, *Newton's Third Rule and the Experimental Argument for Universal Gravity* (Routledge, 2021).
2. Daniel Fleisch, *A Student's Guide to Maxwell's Equations* (Cambridge University Press, 2008).
3. Hendrik Antoon Lorentz et al., *The Principle of Relativity: A Collection of Original Memoirs on the Special and General Theory of Relativity*, trans. W. Perrett and G. B. Jeffery (Dover, 1952).

4. Jonathan Oppenheim and Stephanie Wehner, "The Uncertainty Principle Determines the Nonlocality of Quantum Mechanics," *Science* 330, no. 6007 (November 19, 2010): 1072–1074, https://doi.org/10.1126/science.1192065.
5. Gregory J. Chaitin, "Gödel's Theorem and Information," *International Journal of Theoretical Physics* 21 (1982): 941–954, https://doi.org/10.1007/BF02084159.
6. See Margaret J. Anderson, *Carl Linnaeus: Father of Classification*, rev. ed. (Enslow, 2009) and Ernst Mayr, "Lamarck Revisited," *Journal of the History of Biology* 5 (1972): 55–94, https://doi.org/10.1007/BF02113486.
7. Oswald T. Avery, Colin M. MacLeod, and Maclyn McCarty, "Studies on the Chemical Nature of the Substance Inducing Transformation of Pneumococcal Types," *Journal of Experimental Medicine* 79, no. 2 (February 1, 1944): 137–158, https://doi.org/10.1084/jem.79.2.137.
8. J. D. Watson and F. H. C. Crick, "Molecular Structure of Nucleic Acids: A Structure for Deoxyribose Nucleic Acid," *Nature* 171 (1953): 737–738, https://doi.org/10.1038/171737a0.
9. Howard M. Temin and David Baltimore, "RNA-Directed DNA Synthesis and RNA Tumor Viruses," *Advances in Virus Research* 17 (1972): 129–186, https://doi.org/10.1016/S0065-3527(08)60749-6.
10. Francis S. Collins and Victor A. McKusick, "Implications of the Human Genome Project for Medical Science," *JAMA* 285, no. 5 (2001): 540–544, https://doi.org/10.1001/jama.285.5.540.
11. See John William Draper, *History of the Conflict Between Religion and Science* (D. Appleton, 1874) and Andrew Dickson White, *A History of the Warfare of Science with Theology in Christendom* (Macmillan, 1896).
12. David Hutchings and James C. Ungureanu, *Of Popes and Unicorns: Science, Christianity, and How the Conflict Thesis Fooled the World* (Oxford University Press, 2021), 15–16.
13. Michael Jones, "The Origins of Young Earth Creationism," Peaceful Science, April 11, 2022, https://peacefulscience.org/prints/origns-yec/.
14. "The Evolution of Pew Research Center's Survey Questions about the Origins and Development of Life on Earth," Pew Research Center, February 6, 2019, https://www.pewresearch.org/religion/2019/02/06/the-evolution-of-pew-research-centers-survey-questions-about-the-origins-and-development-of-life-on-earth/.
15. Cary Funk, "How Highly Religious Americans View Evolution Depends on How They're Asked about It," Pew Research Center, February 6, 2019, https://www.pewresearch.org/short-reads/2019/02/06/how-highly-religious-americans-view-evolution-depends-on-how-theyre-asked-about-it/.
16. Terry Godfrey, Twitter direct message to author, March 28, 2024.
17. Nathaniel T. Jeanson and Jason Lisle, "On the Origin of Eukaryotic Species' Genotypic and Phenotypic Diversity," *Answers Research Journal* 9 (2016): 81–122, https://assets.answersresearchjournal.org/doc/v9/genetics-heterozygosity-molecular-clock.pdf.
18. Don Batten, "Ligers and Wholphins? What Next?," *Creation* 22, no. 3 (June 2000): 28–33, https://creation.com/ligers-and-wholphins-what-next.
19. Warren E. Johnson et al., "The Late Miocene Radiation of Modern Felidae: A Genetic Assessment," *Science* 311, no. 5757 (January 6, 2006): 73–77, https://doi.org/10.1126/science.1122277.

20. An interesting article in the creationist *Answers Magazine* is titled "Surprising Similarities Between Creation and Evolution." See John UpChurch, "Surprising Similarities Between Creation and Evolution," *Answers Magazine*, October 8, 2017, https://answersingenesis.org/creation-vs-evolution/surprising-similarities-between-creation-evolution/.
21. Facebook post on a private account, January 5, 2024.
22. David Lack, *Darwin's Finches* (Cambridge University Press, 1983).
23. In her book *The God of Monkey Science: People of Faith in a Modern Scientific World* (Eerdmans, 2023), Janet Kellogg Ray writes about evolution as being part, or even a central component, of the fierce culture war now being waged in the United States and catalogs two other scientific issues that are often accompanied by evolution in discussions related to this warfare—COVID-19 and vaccine denial, and climate change denial. In all three cases, Ray documents the pervading underlying view among a large proportion of self-identified evangelicals that science and scientists cannot be trusted. Further discussion of the political and social roots of these anti-elitist attitudes and the forces behind the culture war are beyond the scope of this book, but it is important to take into account the fact that so many arguments against evolutionary biology are fueled by these factors.
24. Michael J. Reiss and Michael Ruse, *The New Biology: A Battle between Mechanism and Organicism* (Harvard University Press, 2023).
25. Michael J. Reiss and Michael Ruse, "The Battle between Mechanism and Organicism in Biology," *August 2023 Blog*, International Society for Science and Religion, August 15, 2023, https://www.issr.org.uk/blog/august-2023-blog/.

**CHAPTER 2: THE EVOLUTION OF EVOLUTION**

1. Michael Stirrat and R. Elisabeth Cornwell, "Eminent Scientists Reject the Supernatural: A Survey of the Fellows of the Royal Society," *Evolution: Education and Outreach* 6, no. 33 (2013), https://evolution-outreach.biomedcentral.com/articles/10.1186/1936-6434-6-33.
2. Richard Dawkins, *The Blind Watchmaker: Why the Evidence of Evolution Reveals a Universe Without Design* (W. W. Norton, 1996), 6.
3. George Webb, *The Evolution Controversy in America* (University Press of Kentucky, 2002), 19.
4. Richard A. Watson and Eörs Szathmáry, "How Can Evolution Learn?," *Trends in Ecology and Evolution* 31, no. 2 (February 2016): 147–157, https://www.sciencedirect.com/science/article/abs/pii/S0169534715002931.
5. Bendik Hellem Aaby and Hugh Desmond, "Niche Construction and Teleology: Organisms as Agents and Contributors in Ecology, Development, and Evolution," *Biology and Philosophy* 36, no. 47 (September 20, 2021): 1–20, https://doi.org/10.1007/s10539-021-09821-2.
6. William B. Miller Jr., "Biological Information Systems: Evolution as Cognition-Based Information Management," *Progress in Biophysics and Molecular Biology* 134 (May 2018): 1–26, https://doi.org/10.1016/j.pbiomolbio.2017.11.005.
7. František Baluška, Arthur S. Reber, and William B. Miller Jr., "Cellular Sentience as the Primary Source of Biological Order and Evolution," *Biosystems* 218 (August 2022): 104694, https://doi.org/10.1016/j.biosystems.2022.104694.

8. Olen R. Brown and David A. Hullender, "Biological Evolution Requires an Emergent, Self-Organizing Principle," *Progress in Biophysics and Molecular Biology* 182 (September 2023): 75–102, https://doi.org/10.1016/j.pbiomolbio.2023.06.001.
9. Peter A. Corning, "Beyond the Modern Synthesis: A Framework for a More Inclusive Biological Synthesis," *Progress in Biophysics and Molecular Biology* 153 (July 2020): 5–12, https://doi.org/10.1016/j.pbiomolbio.2020.02.002.
10. Perry Marshall, "The Role of Quantum Mechanics in Cognition-Based Evolution," *Progress in Biophysics and Molecular Biology* 180–181 (July–August 2023): 131–139, https://doi.org/10.1016/j.pbiomolbio.2023.04.007.
11. R. A. Fisher, *The Genetical Theory of Natural Selection*, ed. J. H. Bennett (Oxford University Press, 1999).
12. Richard Dawkins, "Selfish Genes and Selfish Memes," in *The Mind's I: Fantasies and Reflections on Self and Soul*, ed. Douglas R. Hofstadter and Daniel C. Dennett (Basic Books, 1981), 124–144.
13. Peter A. Corning et al., eds., *Evolution "On Purpose": Teleonomy in Living Systems* (MIT Press, 2023), 3.
14. Roger Lewin, "Evolutionary Theory under Fire: An Historic Conference in Chicago Challenges the Four-Decade Long Dominance of the Modern Synthesis," *Science* 210, no. 4472 (November 21, 1980): 883–887, https://doi.org/10.1126/science.6107993.
15. Dawkins, *Blind Watchmaker*, 6.
16. Stephen Jay Gould and Niles Eldredge, "Punctuated Equilibria: The Tempo and Mode of Evolution Reconsidered," *Paleobiology* 3, no. 2 (Spring 1977): 115–151, http://mechanism.ucsd.edu/~bill/teaching/philbio/readings/gould.eldridge.punceq.1977.pdf.
17. Motoo Kimura, "The Neutral Theory of Molecular Evolution," *Scientific American*, November 1, 1979, 98, https://www.scientificamerican.com/article/the-neutral-theory-of-molecular-evo/.
18. Richard W. Burkhardt Jr., "Lamarck, Evolution, and the Inheritance of Acquired Characters," *Genetics* 194, no. 4 (August 1, 2013): 793–805, https://doi.org/10.1534/genetics.113.151852.
19. Walter Doerfler, "DNA Methylation and Gene Activity," *Annual Review of Biochemistry* 52 (July 1983): 93–124, https://doi.org/10.1146/annurev.bi.52.070183.000521.
20. Elizabeth J. Radford, "Exploring the Extent and Scope of Epigenetic Inheritance," *Nature Reviews Endocrinology* 14, no. 6 (June 2018): 345–355, https://doi.org/10.1038/s41574-018-0005-5.
21. Simon Conway Morris, *Life's Solution: Inevitable Humans in a Lonely Universe* (Cambridge University Press, 2004).
22. Ron Amundson, "Typology Reconsidered: Two Doctrines on the History of Evolutionary Biology," *Biology and Philosophy* 13, no. 2 (April 1998): 153–177, https://doi.org/10.1023/A:1006599002775.
23. Michael Denton, *Evolution: Still a Theory in Crisis* (Discovery Institute Press, 2016).
24. David Sepkoski, "'Replaying Life's Tape': Simulations, Metaphors, and Historicity in Stephen Jay Gould's View of Life," *Studies in History and Philosophy of Biological and Biomedical Sciences* 58 (August 2016): 73–81, https://doi.org/10.1016/j.shpsc.2015.12.009.

25. Andreas Wagner, *Arrival of the Fittest: Solving Evolution's Greatest Puzzle* (Current, 2014). The title refers to the fact that while natural selection can choose the fittest phenotypes, it cannot produce them.
26. Charles Darwin, *On the Origin of Species: Or; The Preservation of the Favoured Races in the Struggle for Life* (Read & Co. Science, 2022), chap. 5.
27. Andrew Murray, "Salvador Luria and Max Delbrück on Random Mutation and Fluctuation Tests," *Genetics* 202, no. 2 (February 1, 2016): 367–368, https://doi.org/10.1534/genetics.115.186163.
28. John Cairns, Julie Overbaugh, and Stephan Miller, "The Origin of Mutants," *Nature* 335 (1988): 142–145, https://doi.org/10.1038/335142a0.
29. Devon M. Fitzgerald and Susan M. Rosenberg, "What Is Mutation? A Chapter in the Series: How Microbes 'Jeopardize' the Modern Synthesis," *PLOS Genetics* 15, no. 4 (April 1, 2019): e1007995, https://doi.org/10.1371/journal.pgen.1007995.
30. Donald A. Croll et al., "Introduced Predators Transform Subarctic Islands from Grassland to Tundra," *Science* 307, no. 5717 (March 25, 2005): 1959–1961, https://doi.org/10.1126/science.1108485.
31. Clive G. Jones, John H. Lawton, and Moshe Shachak, "Positive and Negative Effects of Organisms as Physical Ecosystem Engineers," *Ecology* 78, no. 7 (October 1997): 1946–1957, https://www.jstor.org/stable/2265935.
32. F. John Odling-Smee, Kevin N. Laland, and Marcus W. Feldman, *Niche Construction: The Neglected Process in Evolution* (Princeton University Press, 2003), 419.
33. Kevin N. Laland et al., "More on How and Why: Cause and Effect in Biology Revisited," *Biology and Philosophy* 28 (2013): 719–745, https://doi.org/10.1007/S10539-012-9335-1.
34. Kenichi Aoki, "A Stochastic Model of Gene-Culture Coevolution Suggested by the 'Culture Historical Hypothesis' for the Evolution of Adult Lactose Absorption in Humans," *Proceedings of the National Academy of Sciences of the United States of America* 83, no. 9 (May 1986): 2929–2933, https://www.pnas.org/doi/10.1073/pnas.83.9.2929.
35. Augusto Anguita-Ruiz, Concepción M. Aguilera, and Ángel Gil, "Genetics of Lactose Intolerance: An Updated Review and Online Interactive World Maps of Phenotype and Genotype Frequencies," *Nutrients* 12, no. 9 (2020): 2689, https://doi.org/10.3390/nu12092689.
36. Isabelle S. Peter and Eric H. Davidson, "Assessing Regulatory Information in Developmental Gene Regulatory Networks," *PNAS* 144, no. 23 (2017): 5862–5869, https://doi.org/10.1073/pnas.1610616114; S. Garte and A. Albert, "Genotype Components as Predictors of Phenotype in Model Gene Regulatory Networks," *Acta Biotheoretica* 67 (2019): 299–320, https://link.springer.com/article/10.1007/s10441-019-09350-2.
37. A leading biologist who has been writing about the ways cells control their own fates as well as their genomes is Denis Noble. In technical papers, articles, and books, Dr. Noble has presented many illustrations of how nongenetic elements of cells and organs play a major role in determining function and even structure in living creatures. One example he points to is the amazing way that the mammalian immune system functions. This remarkable system illustrates a number of features of the cellular control of genetic action that go far beyond the standard "gene X makes

protein X that catalyzes the X metabolic reaction" model. Raymond Noble and Denis Noble, "Was the Watchmaker Blind? Or Was She One-Eyed?," *Biology* 6, no. 4 (2017): 47, https://doi.org/10.3390/biology6040047.

38. James A. Shapiro, "How Life Changes Itself: The Read-Write (RW) Genome," *Physics of Life Reviews* 10, no. 3 (September 2013): 287–323, https://doi.org/10.1016/j.plrev.2013.07.001.
39. William Paley, *Natural Theology: Or Evidence of the Existence and Attributes of the Deity, Collected from the Appearances of Nature*, ed. Matthew D. Eddy and David Knight (Oxford University Press, 2008).
40. Noble and Noble, "Was the Watchmaker Blind?"
41. Daniel C. Dennett, "The Evolution of Reasons," in *Contemporary Philosophical Naturalism and Its Implications*, ed. Bana Bashour and Hans D. Muller (Routledge, 2014), 49.

**CHAPTER 3: THE FUNDAMENTAL PRINCIPLE OF BIOLOGY: INHERITANCE**

1. Charles Darwin, *On the Origin of Species: Or; The Preservation of the Favoured Races in the Struggle for Life* (Read & Co. Science, 2022), chap. 4, emphasis added.
2. Theresa Bullard et al., "Test of Cairns-Smith's 'Crystals-as-Genes' Hypothesis," *Faraday Discussions* 136 (2007): 231–245, https://doi.org/10.1039/B616612C.
3. Granville Sewell, "Human-Engineered Self-Replicating Machines," *BioCosmos* 3, no. 1 (January 2023): 12–15, https://doi.org/10.2478/biocosmos-2023-0002.
4. Douglas Blackiston et al., "A Cellular Platform for the Development of Synthetic Living Machines," *Science Robotics* 6, no. 52 (March 17, 2021): eabf1571, https://doi.org/10.1126/scirobotics.abf1571.
5. Prof. Bricks - LEGO Robotics, "LEGO MINDSTORMS Robot Builds Copies of Itself," posted April 10, 2015, YouTube, 1:06, https://www.youtube.com/watch?v=6siv8HTT814.
6. J. D. Watson and F. H. C. Crick, "Molecular Structure of Nucleic Acids: A Structure for Deoxyribose Nucleic Acid," *Nature* 171 (1953): 737–738, https://doi.org/10.1038/171737a0.
7. Randy Isaac and Stephen Freeland, "Rethinking Abiogenesis: Part III, Meaning in the Light of Complexity, Information, and Simplification," *Perspectives on Science and Christian Faith* 75, no. 1 (March 2023): 16–28, https://doi.org/10.56315/PSCF3-23Isaac.
8. Stephen C. Meyer, *Return of the God Hypothesis: Three Scientific Discoveries That Reveal the Mind Behind the Universe* (HarperOne, 2021); Stephen C. Meyer, "DNA and the Origin of Life: Information, Specification, and Explanation," in *Darwinism, Design and Public Education*, ed. John Angus Campbell and Stephen C. Meyer (Michigan State University Press, 2003), 223–285.
9. Seymour Garte, "Accurate Phenotypic Self-Replication as a Necessary Cause for Biological Evolution," *BioSystems* 237 (March 2024): 105154, https://doi.org/10.1016/j.biosystems.2024.105154.
10. German biochemist Manfred Eigen was the first to write about thresholds needed for accurate replication, and he coined the terms "error threshold" and "error catastrophe." He also showed the relationship between error catastrophe and the "average selective advantage" of a favored allele. See Manfred Eigen and Peter

Schuster, "The Hypercycle: A Principle of Natural Self-Organization. Part B: The Abstract Hypercycle," *Naturwissenschaften* 65 (January 1978): 7–41, https://link.springer.com/article/10.1007/BF00420631.

11. Vitaly Vanchurin et al., "Toward a Theory of Evolution as Multilevel Learning," *Proceedings of the National Academy of Sciences of the United States of America* 119, no. 6 (February 4, 2022): e2120037119, https://doi.org/10.1073/pnas.2120037119.
12. Fred S. Cannon, "Vocabulary about Nature Used by Moses in Genesis–Deuteronomy," lecture, ASA 2023 conference, University of Toronto Mississauga, Ontario, July 31, 2023, posted by American Scientific Affiliation, October 19, 2023, YouTube, 26:15, https://www.youtube.com/watch?v=CBJyA7QkJxA.

## CHAPTER 4: AGENCY, COGNITION, AND TELEOLOGY (ACT) IN BIOLOGY

1. Etienne Gilson, *From Aristotle to Darwin and Back Again: A Journey in Final Causality, Species, and Evolution*, trans. John Lyon (Ignatius Press, 2009; repr., University of Notre Dame Press, 1984), 99.
2. Marjorie Grene, "Aristotle and Modern Biology," *Journal of the History of Ideas* 33, no. 3 (July–September 1972): 395–424, https://doi.org/10.2307/2709043.
3. James A. Shapiro, "All Living Cells Are Cognitive," *Biochemical and Biophysical Research Communications* 564 (July 30, 2021): 134–149, https://doi.org/10.1016/j.bbrc.2020.08.120.
4. František Baluška, William B. Miller Jr., and Arthur S. Reber, "Biomolecular Basis of Cellular Consciousness via Subcellular Nanobrains," *International Journal of Molecular Sciences* 22, no. 5 (March 3, 2021): 2545, https://doi.org/10.3390/ijms22052545.
5. Audrey Dussutour, "Learning in Single Cell Organisms," *Biochemical and Biophysical Research Communications* 564 (July 30, 2021): 92–102, https://doi.org/10.1016/j.bbrc.2021.02.018.
6. Keith D. Farnsworth, "How Biological Codes Break Causal Chains to Enable Autonomy for Organisms," *BioSystems* 232 (October 2023): 105013, https://doi.org/10.1016/j.biosystems.2023.105013.
7. William B. Miller Jr., František Baluška, and Arthur S. Reber, "A Revised Central Dogma for the 21st Century: All Biology Is Cognitive Information Processing," *Progress in Biophysics and Molecular Biology* 182 (September 2023): 34–48, https://doi.org/10.1016/j.pbiomolbio.2023.05.005.
8. While generally attributed to J. B. S. Haldane, it appears this quote may have originated with physiologist Ernst Wilhelm von Brücke in a lecture. See Walter B. Cannon, *The Way of an Investigator: A Scientist's Experiences in Medical Research* (Hafner, 1968), 108.
9. Timothy Lenoir, *The Strategy of Life: Teleology and Mechanics in Nineteenth-Century German Biology* (University of Chicago Press, 1989), ix.
10. Ernst Mayr, *Toward a New Philosophy of Biology: Observations of an Evolutionist* (Harvard University Press, 1988), 39.
11. Luisa Cochella and Rachel Green, "Fidelity in Protein Synthesis," *Current Biology* 15, no. 14 (July 26, 2005): PR536–R540, https://doi.org/10.1016/j.cub.2005.07.018.
12. Bruce Alberts et al., *Molecular Biology of the Cell*, 5th ed. (Garland Science, 2008), 371.

13. Daniel C. Dennett, "The Evolution of Reasons," in *Contemporary Philosophical Naturalism and Its Implications*, ed. Bana Bashour and Hans D. Muller (Routledge, 2014), 49.
14. Raymond Noble and Denis Noble, "Was the Watchmaker Blind? Or Was She One-Eyed?," *Biology* 6, no. 4 (2017): 47, https://doi.org/10.3390/biology6040047.
15. Peter A. Corning et al., eds., *Evolution "On Purpose": Teleonomy in Living Systems* (MIT Press, 2023), 4.
16. Devon M. Fitzgerald and Susan M. Rosenberg, "What Is Mutation? A Chapter in the Series: How Microbes 'Jeopardize' the Modern Synthesis," *PLOS Genetics* 15, no. 4 (April 1, 2019): e1007995, https://doi.org/10.1371/journal.pgen.1007995.
17. Noble and Noble, "Was the Watchmaker Blind?"
18. Sam Harris, *Free Will* (Free Press, 2012).
19. Benjamin Libet et al., "Time of Conscious Intention to Act in Relation to Onset of Cerebral Activity (Readiness-Potential): The Unconscious Initiation of a Freely Voluntary Act," *Brain* 106, no. 3 (September 1983): 623–642, https://doi.org/10.1093/brain/106.3.623.
20. Steve Taylor, *Spiritual Science: Why Science Needs Spirituality to Make Sense of the World* (Watkins, 2018).
21. Steve Taylor, "How a Flawed Experiment 'Proved' That Free Will Doesn't Exist," *Scientific American*, December 6, 2019, https://www.scientificamerican.com/blog/observations/how-a-flawed-experiment-proved-that-free-will-doesnt-exist/.
22. Taylor, "How a Flawed Experiment 'Proved' That Free Will Doesn't Exist."
23. Justin Brierley, *The Surprising Rebirth of Belief in God: Why New Atheism Grew Old and Secular Thinkers Are Considering Christianity Again* (Tyndale Elevate, 2023).
24. Denis Alexander and Alister McGrath, eds., *Coming to Faith Through Dawkins: 12 Essays on the Pathway from New Atheism to Christianity* (Kregel, 2023).
25. Sara J. Shettleworth, *Cognition, Evolution, and Behavior*, 2nd ed. (Oxford University Press, 2010), 4.
26. Pamela Lyon, "The Cognitive Cell: Bacterial Behavior Reconsidered," *Frontiers in Microbiology* 6 (April 13, 2015): 264, https://doi.org/10.3389/fmicb.2015.00264.
27. William B. Miller Jr., Francisco J. Enguita, and Ana Lúcia Leitão, "Non-Random Genome Editing and Natural Cellular Engineering in Cognition-Based Evolution," *Cells* 10, no. 5 (2021): 1125, https://doi.org/10.3390/cells10051125; William B. Miller Jr. et al., "Cellular and Natural Viral Engineering in Cognition-Based Evolution," *Communicative and Integrative Biology* 16, no. 1 (May 2, 2023): 2196145, https://doi.org/10.1080/19420889.2023.2196145.
28. William S. Hlavacek et al., "Rules for Modeling Signal-Transduction Systems," *Science's STKE* 2006, no. 344 (July 18, 2006): re6, https://doi.org/10.1126/stke.3442006re6.
29. James A. Shapiro, "Genome Informatics: The Role of DNA in Cellular Computations," *Biological Theory* 1, no. 3 (2006): 288–301, https://doi.org/10.1162/BIOT.2006.1.3.288.
30. Brian J. Ford, "The Secret Power of the Single Cell," *NewScientist* 206, no. 2757 (April 21, 2010): 26–27, https://doi.org/10.1016/S0262-4079(10)61011-1.
31. Jordi van Gestel, Hera Vlamakis, and Roberto Kolter, "From Cell Differentiation to Cell Collectives: *Bacillus subtilis* Uses Division of Labor to Migrate," *PLOS Biology* 13, no. 4 (April 20, 2015): e1002141, https://doi.org/10.1371/journal.pbio.1002141.

32. Oné R. Pagán, *The First Brain: The Neuroscience of Planarians* (Oxford University Press, 2014), 152.
33. Baluška, Miller, and Reber, "Biomolecular Basis."
34. Fazale Rana has discussed this seemingly odd fact in a recent book titled *Is Darwinian Evolution Left Unexplained?* (Reasons to Believe, 2023), Reasons to Believe, https://get.reasons.org/darwinian-evolution-unexplained-lp/.
35. María Romeralo, Ricardo Escalante, and Sandra L. Baldauf, "Evolution and Diversity of Dictyostelid Social Amoebae," *Protist* 163, no. 3 (May 2012): 327–343, https://doi.org/10.1016/j.protis.2011.09.004.
36. Zoltán Barta, "Individual Variation Behind the Evolution of Cooperation," *Philosophical Transactions of the Royal Society B: Biological Sciences* 371, no. 1687 (February 2016): 20150087, https://doi.org/10.1098/rstb.2015.0087.
37. I have published a paper describing some theoretical aspects of this amazing phenomenon: Seymour Garte, "Targeted Hypermutation as a Survival Strategy: A Theoretical Approach," *Acta Biotheoretica* 71, no. 4 (September 5, 2023): article 20, https://doi.org/10.1007/s10441-023-09471-9.
38. Philip Ball, *How Life Works: A User's Guide to the New Biology* (University of Chicago Press, 2023).
39. The theological implications of the universal presence of purpose in life have been addressed by many, including Christian biologist Denis Alexander in his book *Is There Purpose in Biology?: The Cost of Existence and the God of Love* (Monarch Books, 2018).

**CHAPTER 5: ABIOGENESIS: THE ORIGIN OF LIFE**

1. John D. Sutherland, "Opinion: Studies on the Origin of Life—the End of the Beginning," *Nature Reviews Chemistry* 1, no. 2 (January 18, 2017): 0012, https://doi.org/10.1038/s41570-016-0012.
2. "Bias in the Scientific Community: Dr Tour and Dr Sy Garte Discuss His New Book," Dr. James Tour, *The Science and Faith Podcast*, October 1, 2021, YouTube, 1:21:28, https://www.youtube.com/watch?v=C_neIY8aKn8.
3. Richard Dawkins, *The God Delusion* (Mariner Books, 2008), 167.
4. Daniel C. Dennett, *Darwin's Dangerous Idea: Evolution and the Meanings of Life* (Simon & Schuster, 1996).
5. Rosa Larralde, Michael P. Robertson, and Stanley L. Miller, "Rates of Decomposition of Ribose and Other Sugars: Implications for Chemical Evolution," *Proceedings of the National Academy of Sciences of the United States of America* 92, no. 18 (August 29, 1995): 8158–8160, https://doi.org/10.1073/pnas.92.18.8158.
6. Jeremy England, *Every Life Is on Fire: How Thermodynamics Explains the Origins of Living Things* (Basic Books, 2020).
7. Jennifer A. Doudna and Thomas R. Cech, "The Chemical Repertoire of Natural Ribozymes," *Nature* 418, no. 6894 (July 11, 2002): 222–228, https://doi.org/10.1038/418222a.
8. Charles Wilson and Jack W. Szostak, "*In Vitro* Evolution of a Self-Alkylating Ribozyme," *Nature* 374 (1995): 777–782, https://doi.org/10.1038/374777a0.
9. Wendy K. Johnston et al., "RNA-Catalyzed RNA Polymerization: Accurate and General RNA-Templated Primer Extension," *Science* 292, no. 5520 (May 18, 2001): 1319–1325, https://doi.org/10.1126/science.1060786.

10. Walter L. Bradley, "Information, Entropy, and the Origin of Life," in *Debating Design: From Darwin to DNA*, ed. William A. Dembski and Michael Ruse (Cambridge University Press, 2004), 346.
11. Peter R. Wills and Charles W. Carter Jr., "Insuperable Problems of the Genetic Code Initially Emerging in an RNA World," *Biosystems* 164 (2018): 155–166, https://doi.org/10.1016/j.biosystems.2017.09.006.
12. Wills and Carter Jr., "Insuperable Problems of the Genetic Code."
13. Jack W. Szostak, "The Eightfold Path to Non-Enzymatic RNA Replication," *Journal of Systems Chemistry* 3, no. 1 (February 3, 2012): article 2, http://www.jsystchem.com/content/3/1/2.
14. Charles W. Carter Jr. and Peter R. Wills, "Interdependence, Reflexivity, Fidelity, Impedance Matching, and the Evolution of Genetic Coding," *Molecular Biology and Evolution* 35, no. 2 (February 2018): 269–286, https://doi.org/10.1093/molbev/msx265.
15. Yuri I. Wolf and Eugene V. Koonin, "On the Origin of the Translation System and the Genetic Code in the RNA World by Means of Natural Selection, Exaptation, and Subfunctionalization," *Biology Direct* 2, no. 14 (2007), https://doi.org/10.1186/1745-6150-2-14.
16. Charles W. Carter Jr., "What RNA World? Why a Peptide/RNA Partnership Merits Renewed Experimental Attention," *Life* 5, no. 1 (2015): 294–320, https://doi.org/10.3390/life5010294.
17. See Seymour Garte, "The Continuity Principle and the Evolution of Replication Fidelity," *Acta Biotheoretica* 69, no. 3 (September 2021): 303–318, https://doi.org/10.1007/s10441-020-09399-4 and Sy Garte, "Evidence for Phase Transitions in Replication Fidelity and Survival Probability at the Origin of Life," *BioCosmos* 1, no. 1 (January 2021): 2–10, https://doi.org/10.2478/biocosmos-2021-0002.
18. Stuart M. Marshall et al., "Identifying Molecules as Biosignatures with Assembly Theory and Mass Spectrometry," *Nature Communications* 12, no. 1 (May 24, 2021): 3033, https://doi.org/10.1038/s41467-021-23258-x.
19. Stuart M. Marshall et al., "Formalising the Pathways to Life Using Assembly Spaces," *Entropy* 24, no. 7 (2022): 884, https://doi.org/10.3390/e24070884.
20. Hubert P. Yockey, "A Calculation of the Probability of Spontaneous Biogenesis by Information Theory," *Journal of Theoretical Biology* 67, no. 3 (August 7, 1977): 377–398, https://doi.org/10.1016/0022-5193(77)90044-3.

**CHAPTER 6: HUMAN BEHAVIOR AND CONSCIOUSNESS**

1. Sy Garte, *The Works of His Hands: A Scientist's Journey from Atheism to Faith* (Kregel, 2019), 87–88.
2. Sy Garte, "A New Christian Meets New Atheism," in *Coming to Faith Through Dawkins: 12 Essays on the Pathway from New Atheism to Christianity*, ed. Denis Alexander and Alister McGrath (Kregel, 2023), 8.
3. Damee Choi, Kenji J. Tsuchiya, and Nori Takei, "Interaction Effect of Oxytocin Receptor *(OXTR)* rs53576 Genotype and Maternal Postpartum Depression on Child Behavioural Problems," *Scientific Reports* 9, no. 1 (2019): 7685, https://doi.org/10.1038/s41598-019-44175-6.
4. Sarina M. Rodrigues et al., "Oxytocin Receptor Genetic Variation Relates to Empathy and Stress Reactivity in Humans," *Proceedings of the National Academy of*

*Sciences of the United States of America* 106, no. 50 (December 15, 2009): 21437–21441, https://doi.org/10.1073/pnas.0909579106.

5. Karolina Pierzynowska et al., "Roles of the Oxytocin Receptor (OXTR) in Human Diseases," *International Journal of Molecular Sciences* 24, no. 4 (2023): 3887, https://doi.org/10.3390/ijms24043887.
6. Anya C. Hurlbert and Yazhu Ling, "Biological Components of Sex Differences in Color Preference," *Current Biology* 17, no. 16 (August 21, 2007): R623–R625, https://doi.org/10.1016/j.cub.2007.06.022.
7. Garte, *Works of His Hands*, 119.
8. "What Is Evolutionary Creation?," BioLogos, updated February 6, 2024, https://biologos.org/common-questions/what-is-evolutionary-creation.
9. Thomas H. Huxley, *Lessons in Elementary Physiology* (Macmillan, 1866), 193.
10. Garte, *Works of His Hands*, 112.
11. Roger Penrose, *The Emperor's New Mind: Concerning Computers, Minds and the Laws of Physics* (Oxford University Press, 2016).
12. Lee M. Pierson and Monroe Trout, "What Is Consciousness For?," *New Ideas in Psychology* 47 (December 2017): 62–71, https://doi.org/10.1016/j.newideapsych.2017.05.004.
13. Perry Marshall, *Evolution 2.0: Breaking the Deadlock Between Darwin and Design* (BenBella Books, 2015).
14. Perry Marshall, "Biology Transcends the Limits of Computation," *Progress in Biophysics and Molecular Biology* 165 (October 2021): 88–101, https://doi.org/10.1016/j.pbiomolbio.2021.04.006.
15. Samuel T. Wilkinson, *Purpose: What Evolution and Human Nature Imply about the Meaning of Our Existence* (Pegasus Books, 2024), 9.
16. Wilkinson, 149.
17. Marco Masi, "An Evidence-Based Critical Review of the Mind-Brain Identity Theory," *Frontiers in Psychology* 14 (October 26, 2023): 1150605, https://doi.org/10.3389/fpsyg.2023.1150605.
18. Sharon Hewitt Rawlette, "Why Science Will Never Explain Consciousness," *Psychology Today*, updated July 31, 2024, https://www.psychologytoday.com/us/blog/mysteries-of-consciousness/202407/why-science-will-never-explain-consciousness.
19. A good source, for those who want to dig deeper, is the classic book *Gödel, Escher, Bach: An Eternal Golden Braid* by Douglas R. Hofstadter (Basic Books, 1979).
20. "Roger Penrose: Physics of Consciousness and the Infinite Universe: Lex Fridman Podcast #85," *Lex Fridman Podcast*, March 31, 2020, YouTube, 1:27:56, https://www.youtube.com/watch?v=orMtwOz6Db0&t=584s.
21. Paul M. Gould, *A Good and True Story: Eleven Clues to Understanding Our Universe and Your Place in It* (Brazos Press, 2022).
22. Paul M. Gould, "A Good and True Love," *The Worldview Bulletin Newsletter*, February 5, 2024, https://worldviewbulletin.substack.com/p/a-good-and-true-love.

### CHAPTER 7: DIVINE DESIGN: THE BEGINNING

1. Henry Drummond, *The Lowell Lectures on the Ascent of Man* (James Pott, 1907), 333.
2. Drummond, 334.
3. Michael Guillen, *Believing Is Seeing: A Physicist Explains How Science Shattered His Atheism and Revealed the Necessity of Faith* (Tyndale Refresh, 2021); John

Polkinghorne, *Quantum Physics and Theology: An Unexpected Kinship* (Yale University Press, 2008).

4. Sy Garte, *Science and Faith in Harmony: Contemplations on a Distilled Doxology* (Kregel, 2024), 194–198.
5. See chapter 13 of *Science and Faith in Harmony*, 75–79.
6. Jim Stump, *The Sacred Chain: How Understanding Evolution Leads to Deeper Faith* (HarperOne, 2024), 244.
7. Simon Conway Morris, *Life's Solution: Inevitable Humans in a Lonely Universe* (Cambridge University Press, 2004).
8. James Ussher, *The Annals of the World. Deduced from The Origin of Time, and Continued to the Beginning of the Emperor Vespasian's Reign [. . .]*, 1658.

**APPENDIX: TECHNICAL DETAILS**

1. J. D. Watson and F. H. C. Crick, "Molecular Structure of Nucleic Acids: A Structure for Deoxyribose Nucleic Acid," *Nature* 171 (1953): 737–738, https://doi.org/10.1038/171737a0.
2. Michael M. Hoffman et al., "AANT: The Amino Acid–Nucleotide Interaction Database," *Nucleic Acids Research* 32, Database issue (2004): D174–D181, https://doi.org/10.1093/nar/gkh128.
3. Bruce Alberts et al., *Molecular Biology of the Cell*, 5th ed. (Garland Science; 2008), 367–382.
4. See Sy Garte, "Protein Synthesis," March 27, 2022, YouTube, 15:21, https://www.youtube.com/watch?v=GgilYN5a_9A&t=95s and Sy Garte, "BioChemist Argues God Exists . . . Using Biology, Dr. Sy Garte: CCv1 Session 3," by Capturing Christianity, January 19, 2022, YouTube, 46:36, https://www.youtube.com/watch?v=Hw7DG7L6Gsw.
5. Bruce Alberts et al., *Molecular Biology of the Cell*, 5th ed. (Garland Science, 2008).
6. Jocelyn E. Krebs, Elliott S. Goldstein, and Stephen T. Kilpatrick, *Lewin's Genes XII* (Jones & Bartlett Learning, 2017).

# About the Author

SEYMOUR "SY" GARTE, PHD (in biochemistry), is the author of the award-winning book *The Works of His Hands: A Scientist's Journey from Atheism to Faith* (Kregel, 2019) and the bestselling *Science and Faith in Harmony: Contemplations on a Distilled Doxology* (Kregel, 2024).

He has been a tenured professor at New York University, Rutgers University, and the University of Pittsburgh; division director at the Center for Scientific Review of the National Institutes of Health; and interim vice president for research at Uniformed Services University of the Health Sciences.

He has published more than two hundred peer-reviewed scientific papers and five books, as well as articles on science and Christian faith in periodicals such as *Christianity Today*, *Premier Christianity*, and *Perspectives on Science and Christian Faith*. He has also appeared on numerous podcasts and YouTube channels, with more than four million views.

Currently, Dr. Garte serves as editor-in-chief of the American Scientific Affiliation's (ASA) online quarterly magazine, *God and Nature*, and also as vice president for the Washington, DC, metro chapter of the ASA. He also served as a member of the board of advisors of the

John Templeton Foundation. Dr. Garte converted to Christianity from an atheist family background, and he is now a certified lay servant in the United Methodist Church in Rockville, Maryland.